WASTEWATER TREATMENT
Troubleshooting and Problem Solving

Glenn M. Tillman
Director of Utilities
Town of Berryville, Virginia

LEWIS PUBLISHERS
Boca Raton London New York Washington, D.C.

Library of Congress Cataloging-in-Publication Data

Tillman, Glenn M.
　　Water treatment : troubleshooting and problem solving / Glenn M. Tillman.
　　　　p. cm.
　　Includes bibliographical references.
　　ISBN 1-57504-000-X
　　1. Sewage disposal plants—Maintenance and repair.. I. Title.
TD746.T55 1996
628.3—dc20　　　　　　　　　　　　　　　　　　　　　　　　　　　　95-34804
　　　　　　　　　　　　　　　　　　　　　　　　　　　　　　　　　　　CIP

　　This book contains information obtained from authentic and highly regarded sources. Reprinted material is quoted with permission, and sources are indicated. A wide variety of references are listed. Reasonable efforts have been made to publish reliable data and information, but the author and the publisher cannot assume responsibility for the validity of all materials or for the consequences of their use.
　　Neither this book nor any part may be reproduced or transmitted in any form or by any means, electronic or mechanical, including photocopying, microfilming, and recording, or by any information storage or retrieval system, without prior permission in writing from the publisher.
　　The consent of CRC Press LLC does not extend to copying for general distribution, for promotion, for creating new works, or for resale. Specific permission must be obtained in writing from CRC Press LLC for such copying.
　　Direct all inquiries to CRC Press LLC, 2000 N.W. Corporate Blvd., Boca Raton, Florida 33431.

　　Trademark Notice: Product or corporate names may be trademarks or registered trademarks, and are used only for identification and explanation, without intent to infringe.

© 1996 by CRC Press LLC
Lewis Publishers is an imprint of CRC Press LLC

No claim to original U.S. Government works
International Standard Book Number 1-57504-000-X
Library of Congress Card Number 95-34804
Printed in the United States of America　　1 2 3 4 5 6 7 8 9 0
Printed on acid-free paper

Dedicated to our Father, maker of heaven and earth, and to Jesus Christ, His only Son, our Lord, the only true troubleshooter and problem solver.

And there are diversities of operations, but it is the same God which worketh all in all.

1 *Corinthians* 12:6

ACKNOWLEDGMENTS

I would like to acknowledge and thank Brenda Cunningham, Deborah Knight, and Sandra Woodward for word processing and William S. M'Coy for review comments. Their ready assistance throughout this project has been appreciated.

NOTE

Neither the author nor Ann Arbor Press is responsible for any personal injury or equipment damage caused by the misapplication of the procedures described in this book.

AUTHOR

Glenn M. Tillman, Director of Utilities for the Town of Berryville, Virginia, was formerly an Operations and Maintenance Specialist with Malcolm Pirnie, Inc., a full-service environmental engineering company. As a consultant, Mr. Tillman prepared operation and maintenance manuals, developed "how to" operator's guidebooks, wrote standard operating procedures, compiled operator training curricula, instructed plant personnel, and conducted state operator certification preparation programs. He was responsible for the startup, initial operation, and management of treatment plants throughout the United States, and has optimized plant performance by means of alternative operating procedures, performing on-site operations diagnostic assessments, preventive maintenance management, and time analysis staffing evaluations.

Mr. Tillman has over 15 years of operations, management, maintenance, training, and consulting experience at municipal and industrial water and wastewater facilities. He is a certified environmental operator in both water and wastewater treatment.

Mr. Tillman received an A.S. degree in biology from Germanna Community College in 1976, and a B.A. degree in religion and geology from Mary Washington College in 1979, where he was graduated summa cum laude and was elected to Phi Beta Kappa.

PREFACE

This operator's guide presents basic troubleshooting and problem solving information for typical problems that can occur during the operation of processes used at municipal and/or industrial wastewater treatment plants. Common problems and the recommended operator responses to try first are listed in tabular form for individual unit processes. If these suggestions do not solve your problem, or if your process or mechanical problem is more unusual or exotic than those covered here, contact your equipment supplier, state engineering staff, or consult one or more of the references listed in this guide's bibliography.

The troubleshooting and problem solving information included is not meant to be a complete, comprehensive solution source for every conceivable problem. The intent here is to provide helpful suggestions for the correction of the most common problems. Entry level operators are the primary audience. More experienced plant operators and managers may also find the information useful as a place to start when faced with a troubleshooting situation.

Only the more commonly encountered wastewater treatment unit processes are included. After an overview of the treatment objective of the unit process, each process is provided with a troubleshooting table divided into Indicators/Observations; Possible Cause; Check or Monitor; Possible Solutions columns.

Piping, pumps, valves, and electrical equipment are not covered separately in individual chapters. However, these components are covered where appropriate in individual unit process chapters, and in a concluding chapter devoted to basic mechanical problems.

The information compiled in this volume has been collected from various equipment manufacturers' operation and maintenance manuals, U.S. Environmental Protection Agency (EPA) technology transfer documents, my work as a plant Operations and Maintenance (O&M) manual writer, and my own experience as a plant manager and operator. Although most of the guidelines provided here are standard accepted industry procedures, you may not always find the recommendations effective. If you discover a better or new way to solve the problem, or have a problem and solution specific to your plant that is not listed here, add it in the "Notes" section at the end of each chapter.

TABLE OF CONTENTS

1.0. USER'S GUIDE 1

2.0. GENERAL TROUBLESHOOTING PROCEDURES 3
    2.1. Introduction 3
    2.2. How to Troubleshoot 3
    2.3. Basic Troubleshooting Skills 4
    2.4. Equipment Lockout 4
    2.5. Removing Lockout 5

3.0. SCREENING 7
    3.1. Process Overview 7
    3.2. Screening Troubleshooting 8

4.0. SHREDDING AND GRINDING 15
    4.1. Process Overview 15
    4.2. Screening and Grinding Troubleshooting 16

5.0. GRIT REMOVAL 19
    5.1. Process Overview 19
    5.2. Grit Removal Troubleshooting 20

6.0. PRIMARY CLARIFICATION 25
    6.1. Process Overview 25
    6.2. Primary Clarification Troubleshooting 27

7.0. LAGOONS/PONDS 39
    7.1. Process Overview 39
    7.2. Lagoon/Pond Troubleshooting 41

8.0. TRICKLING FILTERS 51
    8.1. Process Overview 51
    8.2. Trickling Filter Troubleshooting 53

9.0. ROTATING BIOLOGICAL CONTRACTORS (RBCs) 63
    9.1. Process Overview 63
    9.2. Rotating Biological Contractors Troubleshooting 64

- **10.0. ACTIVATED SLUDGE** 69
  - 10.1. Process Overview 69
  - 10.2. Activated Biofilter (ABF) Process Overview 70
  - 10.3. Activated Sludge Troubleshooting 71
- **11.0. SECONDARY CLARIFIERS** 81
  - 11.1. Process Overview 81
  - 11.2. Secondary Clarifier Troubleshooting 83
- **12.0. NITRIFICATION** 91
  - 12.1. Process Overview 91
  - 12.2. Nitrification Troubleshooting 92
- **13.0. DENITRIFICATION** 97
  - 13.1. Process Overview 97
  - 13.2. Denitrification Troubleshooting 98
- **14.0. EFFLUENT FILTRATION** 101
  - 14.1. Process Overview 101
  - 14.2. Effluent Filtration Troubleshooting 102
- **15.0. CARBON ADSORPTION** 109
  - 15.1. Process Overview 109
  - 15.2. Carbon Adsorption Troubleshooting 110
- **16.0. CHLORINATION** 113
  - 16.1. Process Overview 113
  - 16.2. Chemical Dechlorination Process Overview 113
  - 16.3. Chlorination Troubleshooting 114
  - 16.4. Chlorine Contact Chamber Troubleshooting 121
- **17.0. GRAVITY THICKENING** 125
  - 17.1. Process Overview 125
  - 17.2. Gravity Thickening Troubleshooting 126
- **18.0. DISSOLVED AIR FLOTATION (DAF) THICKENING** 131
  - 18.1. Process Overview 131
  - 18.2. Dissolved air Flotation Thickening Troubleshooting 132
- **19.0. ANAEROBIC DIGESTION** 137
  - 19.1. Process Overview 137
  - 19.2. Anaerobic Digestion Troubleshooting 139

20.0.	AEROBIC DIGESTION	157
	20.1. Process Overview	157
	20.2. Aerobic Digestion Troubleshooting	158
21.0.	CENTRIFUGATION	161
	21.1. Process Overview	161
	21.2. Centrifugation Troubleshooting	162
22.0.	VACUUM FILTRATION	169
	22.1. Process Overview	169
	22.2. Vacuum Filtration Troubleshooting	170
23.0.	PRESSURE FILTRATION	175
	23.1. Process Overview	175
	23.2. Pressure Filtration Troubleshooting	176
24.0.	BELT FILTRATION	181
	24.1. Process Overview	181
	24.2. Belt Filtration Troubleshooting	182
25.0.	BASIC MECHANICAL PROBLEMS	185
	25.1. Basic Mechanical Troubleshooting	186
BIBLIOGRAPHY		201

1 USER'S GUIDE

This operator's guide has been developed as a first source for troubleshooting and solving treatment problems commonly encountered in the daily operation of municipal and industrial wastewater treatment plants. The material is presented in a brief, straightforward format and in easy-to-follow tables to improve the guide's accessibility as a problem solving reference. There is less explanation of how a particular unit process is supposed to work and more on what to do if it is not working properly. Information has been compressed and concentrated with this in mind.

The following chapter covers general troubleshooting procedures appropriate for any unit process at any plant. The chapter provides information on how to troubleshoot, what to look for, and how to correctly lock out equipment to be inspected or worked on. All of the remaining chapters in the guide are devoted to troubleshooting information appropriate for solving problems with individual unit processes commonly encountered at typical wastewater treatment plants.

2 GENERAL TROUBLESHOOTING PROCEDURES

2.1. INTRODUCTION

There are a wide variety of common operational and mechanical problems that can occur periodically to prevent the proper processing of wastes by a treatment plant. Each section of this handbook covers a unit process, the different problems that could occur, and how to correct them. Before this specific guidance can be followed, however, an operator should be familiar with how to "smoke out" or correctly identify the problem.

2.2. HOW TO TROUBLESHOOT

The first step in troubleshooting is to correctly identify the problem. This involves a physical inspection of the equipment or the unit process performance. The problem may be obvious or may require careful investigation. Once identified, monitoring, analyses, and/or observation should be performed before an informed decision can be made as to which corrective measures should be used. In some cases, this data collecting step can be a simple visual observation. In other cases this step may require sampling and laboratory procedures to determine performance efficiency. All information collected should be carefully reviewed in deciding the correct solution to the problem.

The problems covered in this guidebook are those which occur frequently at municipal and industrial wastewater treatment facilities. The solutions suggested are common practice in the industry and are usually successful. There are problems, however, that can resist correction or do not fit into the categories listed here. These may require expert advice from equipment manufacturers or consultation with engineers.

If any of the troubleshooting guidance is found to be ineffective, modify the guidance to fix the problem using your own observations, insight, and experience. Remember that it is the operator's ability to blend experience with technical advice that determines the successful performance of any treatment plant.

2.3. BASIC TROUBLESHOOTING SKILLS

Before investigating the cause of any equipment or process problem, check all the basic operating conditions. Verify that:

- The circuit breaker is closed (on).

- Operating switches are in their proper position.

- For pumps, all needed suction and discharge valves are open, and seal water is available.

- There are no obvious signs of failure, such as jammed, loose, broken, or burned equipment.

- Mounting bolts are tight; couplings, drive bolts, and chains are aligned and correctly tensioned.

Remember to always look for the simplest, most obvious correction to the problem first. If there are no obvious problems, attempt to restart. If the equipment circuit breaker immediately opens after starting, suspect an electrical short (an "overcurrent" condition).

If the equipment runs for a short time but then stops or displays an equipment "fail" alarm, suspect an overload condition (the motor is overloaded and draws more current than is safe).

Remember that accurate troubleshooting is as much an art as a science. A good operator develops his senses to constantly be on the lookout for problems. Operators learn how to "smoke out" the cause of process and equipment problems. Problems may be seen, heard, smelled, or even felt. Troubleshooting often involves an operator's developed intuition regarding what went wrong and how to fix it.

2.4. EQUIPMENT LOCKOUT

Before you work on any mechanical equipment you are troubleshooting, be sure that it is locked out to protect yourself and the safety of other workers. Equipment lockout involves the physical opening (disconnection) of that equipment's electrical circuit breaker and the steps taken to prevent the accidental closing (reconnection) of the circuit breaker while the equipment is being worked on.

To ensure that this equipment lockout has occurred, a lockout procedure should be performed by the personnel who will be performing the work. The following general procedures are offered as a safe, logical lockout sequence to follow:

Step 1

- Notify your co-workers that a specific piece of equipment is being locked out.
- If the equipment is in service, shut the equipment down.

Step 2 - At the Equipment Motor Control Center

- Locate the equipment's circuit breaker.
- Open (disconnect) the correct circuit breaker.
- Attach an "OUT-OF-SERVICE" tag to the open circuit breaker. Fill in information on why and by whom the equipment was taken out of service.
- Physically lock the circuit breaker open, if possible, to further ensure the safety of plant personnel.
- Dissipate any residual energy that may be "stored" in the equipment.
- Check the previous steps and try to start the equipment to verify that it will not operate.

2.5. REMOVING LOCKOUT

After mechanical equipment has been inspected or worked on, the following general guidelines can be used as a correct sequence for removing equipment from lockout.

Step 1 - At the Equipment

- Notify your co-workers that the equipment is being taken out of lockout status.
- Remove any tools or parts from the immediate area of the equipment.

- Ensure that all guards and safety devices have been reinstalled.

- Be sure that all employees are completely clear of the equipment.

Step 2 - At the Motor Control Center

- Remove any locks from the circuit breaker.

- Close the correct equipment circuit breaker.

- Start the equipment.

- If the equipment does not operate properly, lockout the equipment again and arrange for further repairs. Verify that an OUT-OF-SERVICE tag is still in place.

Step 3

- If the equipment operates properly, remove the OUT-OF-SERVICE tag.

- Verify that the equipment status has been entered into the daily log.

The lockout procedures described are adapted from Occupational Safety and Health Administration (OSHA) lockout/tagout rules. The specific rule, Standard 29 CFR Part 1910.147, must be adhered to in detail whenever a guard or other safety device is removed, or whenever you reach into locked out equipment to perform a repair. Minor changes and adjustments on equipment that occur during normal operation are not strictly obligated to follow the OSHA guidelines, but measures should still be taken to provide effective protection.

You are encouraged to obtain a copy of the OSHA standard if you have any questions about correct lockout procedures. It is available from the U.S. Department of Labor, Occupational Safety and Health Administration, Room N3101, Washington, D.C. 20210; telephone (202) 523-9667.

3 SCREENING

3.1. PROCESS OVERVIEW

- Wastewater flows through a screen bar rack to remove large floating objects that could clog pumps or pipes.

- Coarse screens or bar racks are usually made of parallel iron or steel bars spaced 2 to 4 inches apart. Fine screens are spaced 1/2 to 2 inches apart.

- Screens are usually placed in a chamber or channel at a slanted angle position to allow easier cleaning.

- Floating material caught on the screens or racks is removed with rakes either mechanically (automatically) or manually.

- Fine screens may use a travelling screen with a spray water system to remove screenings into a trough.

- Flow velocity is adjusted to 2 to 4 ft/sec for best performance.

- Screens should be cleaned frequently enough to keep water from backing up in the channel leading to the screen which could cause flow surges.

3.2. SCREENING TROUBLESHOOTING

Indicators/Observations	Possible Cause	Check or Monitor	Possible Solutions
1. Odors and/or insects.	1. Accumulation of screened materials.	1. Method and frequency of debris removal. Check hydrogen sulfide levels.	1. Remove accumulated debris more frequently.
2. Too much grit in bar screen chamber. May also cause surging in chamber due to increase in the water level.	2. Flow velocity too low in the screening channel.	2. Flow velocity and water depth to determine amount of grit accumulation.	2. Increase flow velocity (remove screenings more frequently), use fewer screens, or flush regularly with a high pressure hose.
3. Screen clogs often. Wastewater may back up in the screening channel.	3a. Too much debris in the wastewater.	3a. Screen size openings and flow velocity.	3a1. Check for discharge from an industry that could be contributing extra debris. Reduce, control, or remove this added debris source.
			3a2. Use a coarser screen.
	3b. Inadequate cleaning frequency.	3b. Timer or control setting.	3b. Decrease off time. Clean screen more often.

SCREENING

Indicators/Observations	Possible Cause	Check or Monitor	Possible Solutions
4. Mechanical rake inoperative, circuit breaker will not reset.	4. Jammed mechanism.	4. Screen for an obstruction.	4. Remove obstruction.
5. Mechanical rake inoperative, but motor runs.	5a. Broken drive chain.	5a. Inspect drive chain.	5a. Replace or repair chain.
	5b. Broken sheer pin.	5b. Inspect sheer pin.	5b. Replace sheer pin.
	5c. Broken limit switch.	5c. Inspect limit switch.	5c. Replace limit switch.
6. Mechanical rake inoperative, no visible problem.	6a. Failed remote control circuit.	6a. Check switching circuit.	6a. Repair or replace switching circuits.
	6b. Failed motor.	6b. Check motor.	6b. Replace motor.
7. Mechanical screen cycling almost continuously.	7. Overloading of screen.	7. Number of screens in operation and flow rates to screens.	7. Place additional screens in service.
8. Chain clings to sprocket.	8. Material buildup.	8. Tooth pockets.	8. Remove buildup.

Indicators/Observations	Possible Cause	Check or Monitor	Possible Solutions
9. Excessive noise.	9a. Misalignment of sprockets.	9a. Check alignment.	9a. Adjust centers.
	9b. Too little or too much slack.	9b. Slack.	9b. Adjust idler take up.
	9c. Inadequate lubrication.	9c. Lubricating mechanism or schedule.	9c. Clean and re-lubricate.
	9d. Loose components.	9d. Bolts.	9d. Tighten if loose.
	9e. Chain or sprocket worn out.	9e. Wear.	9e. Replace or reverse.
	9f. Too large a chain.	9f. Manufacturer's recommendation.	9f. Replace.
10. Broken torque overload mechanism.	10a. Heavy shock load.	10a. Screen openings.	10a. Manually remove blockage.
	10b. Inadequate lubrication.	10b. Lubricating mechanism or schedule.	10b. Clean and re-lubricate.
	10c. Chain or sprocket corrosion.	10c. Lubrication.	10c. Clean. Protect from corrosion (lubricate).
	10d. Misalignment.	10d. Check moving parts.	10d. Adjust centers.

SCREENING

Indicators/Observations	Possible Cause	Check or Monitor	Possible Solutions
11. Screen binding.	11. Screen travel path out of alignment.	11. Observe screen performance.	11. Adjust screen tracking path.
12. Rake jerks on descent.	12a. Loose rake chain.	12a. Chain tension and slack.	12a. Adjust chain tension.
	12b. Dry roller bearings.	12b. Lubrication schedule.	12b. Lubricate chain rollers and guide rollers.
	12c. Roller guide dirty.	12c. Roller guides.	12c. Clean guides.
13. Screen still clogged after cleaning cycle.	13a. Rake not run often enough.	13a. Rake frequency.	13a. Increase frequency.
	13b. Rake teeth not entering bar spaces.	13b. Rake position on bar.	13b. Adjust rake position.
14. Rake does not stop after completing cycle.	14a. Limit switch arm is out of position.	14a. Limit switch.	14a. Adjust or replace limit switch arm.
	14b. Limit switch is inoperable.	14b. Limit switch.	14b. Replace limit switch.

Indicators/Observations	Possible Cause	Check or Monitor	Possible Solutions
15. Unit shuts down when rake is in channel.	15a. Loose rake chain.	15a. Chain tension and slack.	15a. Adjust chain tension.
	15b. Rake jammed.	15b. Rake movement.	15b. Remove cause of jam.
16. Rake continues to move after unit is shut off.	16. Motor brake not operating.	16. Motor brake.	16. Adjust or replace motor brake.
17. Rake does not track evenly.	17a. Bent rake arms.	17a. Rake arms.	17a. Shim if possible or replace arms.
	17b. Chain tension not even.	17b. Chain tension.	17b. Adjust chain tension.
	17c. Rake arm out of adjustment.	17c. Rake arm.	17c. Adjust arm position.

NOTES

 # SHREDDING AND GRINDING

4.1. PROCESS OVERVIEW

- Shredders and grinders are used to reduce the size of material in the wastewater stream before treatment.

- A comminutor may be used by some plants to both screen and grind material. (Material shredded by a comminutor is returned to the wastewater stream).

- In some shredder/grinder units, all the wastewater flows through the unit(s) and all material is ground automatically.

- In barminutors, shredder/grinder units are combined with a coarse bar screen. In this type, the shredder/grinder may be located near the screen and grind large material diverted to it, or the shredder/grinder unit may be located on the screen itself and slides up and down with the rakes.

4.2. SCREENING AND GRINDING TROUBLESHOOTING

Indicators/Observations	Possible Cause	Check or Monitor	Possible Solutions
1. Equipment inoperative, circuit breaker will not reset.	1. Jammed mechanism.	1. Check equipment for an obstruction.	1. Remove obstruction.
2. Equipment inoperative, motor still runs.	2. Broken coupling.	2. Check coupling.	2. Replace coupling.
3. Receiving chute clogged.	3a. Not enough wash water.	3a. Check wash water flow rate.	3a. Increase wash water flow rate.
	3b. Broken swing hammer blades.	3b. Inspect blades.	3b. Remove and replace broken blades.
4. Larger chunks than usual in the discharge. Pumps may plug from rags, etc.	4a. Dull, worn blades.	4a. Check blades.	4a. Sharpen or replace blades.
	4b. Broken teeth.	4b. Inspect teeth.	4b. Replace any broken teeth.

Note: Visually check size of ground material daily.

NOTES

5 GRIT REMOVAL

5.1. PROCESS OVERVIEW

- Grit chambers remove sand, grit, cinders, small stones, and other similar size and weight material.

- Grit removal equipment can be velocity controlled, aerated, or of the cyclone degritter design.

- Velocity controlled grit chambers have either flow controlled square or rectangular chambers that can be either manually or mechanically cleaned. A chain and flight mechanism or manual cleaning efforts move the grit into a sump or bucket for disposal. Flow velocity should be about 1 ft/sec.

- Aerated grit removal systems inject air into a chamber to produce a spiral flow. Air flow is adjusted to provide a spiral movement while keeping the velocity low enough to allow grit to settle out.

- Cyclone degritters use centrifuge force in a cone shaped unit to remove grit from the wastewater. Wastewater flows into the upper end of the cone to create a vortex that forces the grit particles to the outside of the rotating flow stream. The grit stream produced falls into a grit washer while the degritted flow leaves the cyclone unit through an opening near the top of the unit.

- Grit removed is usually washed to remove organic material prior to its disposal.

5.2. GRIT REMOVAL TROUBLESHOOTING

Indicators/Observations	Possible Cause	Check or Monitor	Possible Solutions
1. Grit packed on collectors.	1a. Collector operating at too great a speed. 1b. Removal equipment operating at too low a speed.	1a. Check collector speed. 1b. Check removal equipment speed.	1a. Reduce collector speed. 1b. Increase removal speed.
2. Cyclone degritter vibration.	2a. Obstruction in the upper port. 2b. Obstruction in the lower port.	2a. Check flow from lower end. 2b. Check flow from lower port.	2a. Reduce flow. 2b. Remove any obstructions.
3. Rotten egg odor (hydrogen sulfide) in grit chamber. Corrosion of metal or concrete.	3. Hydrogen sulfide formation.	3. Sample for total and dissolved sulfides. Check volatile solids content of the grit. Check for floating solids in the grit channel.	3. Wash down chamber and channel and dose with a mild chlorine solution.
4. Accumulated grit in chamber or channels.	4a. Submerged grit. 4b. Flow velocity too low. Broken flight or chain.	4a. Measure depth of grit in chamber. Check for submerged debris on screen. 4b. Check flow velocity. Check equipment.	4a. Wash chamber. Remove collected grit or debris. 4b. Increase flow velocity. Repair equipment.

GRIT REMOVAL

Indicators/Observations	Possible Cause	Check or Monitor	Possible Solutions
5. Corrosion of metal or concrete.	5. Inadequate ventilation (hydrogen sulfide).	5. Ventilation. Check for sludge deposits. Sample for total and dissolved sulfides.	5. Increase or improve ventilation. Install an odor scrubber.
6. Removed grit is grey in color, smells and feels greasy.	6a. Incorrect pressure on cyclone degritter.	6a. Discharge pressure on cyclone degritter.	6a. Keep pressure between 4 and 6 psi by adjusting pump speed.
	6b. Incorrect air flow rate.	6b. Check air flow rate.	6b. Increase air flow rate.
	6c. Grit channel velocity too low.	6c. Use dye or floating objects to determine velocity.	6c. Increase velocity in grit channel.
7. Surface turbulence in aerated grit chamber is reduced.	7. Diffusers covered by rags or grit. Pressure relief valve opened.	7. Diffusers.	7. Clean diffusers. Check screens or other pretreatment processes.
8. Low recovery rate of grit.	8a. Bottom scour.	8a. Velocity.	8a. Adjust velocity to near 1 ft/sec.
	8b. Overaeration.	8b. Aeration.	8b. Reduce aeration.
	8c. Not enough detention time.	8c. Detention time.	8c. Increase detention time by using more units or reducing flow to unit.

Indicators/Observations	Possible Cause	Check or Monitor	Possible Solutions
9. Overflowing grit chamber.	9. Pump surging.	9. Pumps.	9. Adjust pump controls. Control inflow and infiltration.
10. Septic waste with grease and gas bubbles.	10a. Sludge on bottom of chamber.	10a. Grit chamber bottom.	10a. Wash chamber daily. Remove debris.
	10b. Broken collector equipment.	10b. Equipment.	10b. Repair equipment.
11. Removed grit contains excessive organics (aerated grit chambers).	11. Inadequate air flow.	11. Surface turbulence.	11. Increase air flow.

NOTES

6 PRIMARY CLARIFICATION

6.1. PROCESS OVERVIEW

- Primary treatment reduces the organic loading on downstream treatment processes by removing a large amount of settleable, suspended, and floatable materials.

- Primary treatment reduces the velocity of the wastewater through a clarifier to approximately 1 to 2 ft/min so that settling and floatation can take place. This slowing of the flow enhances the removal of the suspended solids in the wastewater.

- Primary settling tanks remove the floated grease and scum, remove the settled sludge solids, and collect them for pumped transfer to disposal or further treatment.

- Clarifiers used may be rectangular or circular. In rectangular clarifiers, the wastewater flows from one end to the other and the settled sludge is moved to a hopper at one end, either by flights set on parallel chains, or by a single bottom scraper set on a travelling bridge. Floating material such as grease and oil are collected by a surface skimmer.

- In circular tanks, the wastewater usually enters at the middle and flows outward. Settled sludge is pushed to a hopper that is in the middle of the tank bottom, and floating material is removed by a surface skimmer.

- Factors affecting primary clarifier performance include the following:

    - Rate of flow through the clarifer.

    - Wastewater characteristics (strength; temperature; amount and type of industrial wastes; and the density, size, and shapes of particles).

- Performance of pretreatment processes.
- Nature and amount of any wastes recycled to the primary clarifier.

■ Key factors in primary clarifier operation include the following concepts:

- Retention time, hours $= \dfrac{(\text{volume, gals})\,(24\text{ hrs/day})}{\text{flow, gallons per day}}$

- Surface loading rate, $\dfrac{\text{gal/day}}{\text{sq/ft}} = \dfrac{\text{flow, gallons per day}}{\text{surface area, sq/ft}}$

- Solids loading rate, $\dfrac{\text{lbs/day}}{\text{sq/ft}} = \dfrac{\text{solids into clarifier, lbs/day}}{\text{surface area, sq/ft}}$

- Weir overflow rate, $\dfrac{\text{gal/day}}{\text{lineal ft}} = \dfrac{\text{flow, gallons per day}}{\text{weir length, lineal ft}}$

6.2. PRIMARY CLARIFICATION TROUBLESHOOTING

Indicators/Observations	Possible Cause	Check or Monitor	Possible Solutions
1. Poor suspended solids removal.	1a. Hydraulic overload.	1a. Flow.	1a. Use all available tanks, choke back flows to increase detention time, or consider adding chemical coagulatants.
	1b. Sludge buildup in tanks reduces volume and allows solids to scour out of tanks.	1b. Pumping duration and sludge levels.	1b. Monitor sludge levels and establish a more frequent and consistent pumping schedule.
	1c. Strong recycle flows.	1c. Quality and quantity of recycle.	1c. Recycle at low flow periods or re-route recycle flows to an alternate process.
	1d. Industrial waste contributions.	1d. Influent sampling.	1d. Identify and restrict any industrial contaminant that may be adversely affecting primary settling.
	1e. Wind currents.	1e. Wind direction.	1e. Install wind barriers.
	1f. Temperature currents.	1f. Wastewater temperature.	1f. Monitor temperature and eliminate storm flow.

PRIMARY CLARIFICATION   27

WASTEWATER TREATMENT TROUBLESHOOTING AND PROBLEM SOLVING

Indicators/Observations	Possible Cause	Check or Monitor	Possible Solutions
2. Sludge washout from tank.	2. Pump cycling. Flow surge.	2. Pump cycling.	2. Modify pumping strategies. Increase tank volume or use an equalization basin.
3. Short circuiting flow.	3a. Uneven weir settings.	3a. Weir settings.	3a. Level weirs.
	3b. Damaged or missing baffles.	3b. Baffles.	3b. Repair, replace, or install baffles.
4. Skimmings build-up.	4a. Skimmings collection system malfunction.	4a. Skimming equipment.	4a. Check skimmings operation and correct if necessary.
	4b. Skimmings not removed often enough.	4b. Removal rate.	4b. Remove skimmings and operate skimming equipment more often.
	4c. Unusual amount of floatable material has entered plant.	4c. Industrial contributors.	4c. Operate skimming equipment as often as possible to remove excess skimmings. Remove material manually, if necessary.

Indicators/Observations	Possible Cause	Check or Monitor	Possible Solutions
5. Skimmings/scum overflow.	5a. Not removed often enough.	5a. Removal rate.	5a. Increase removal rate.
	5b. Heavy industrial waste contributions.	5b. Influent waste.	5b. Identify and restrict industrial contributors.
	5c. Worn or damaged skimmers.	5c. Skimmers.	5c. Clean, repair, or replace skimmers.
	5d. Improper alignment.	5d. Alignment.	5d. Adjust alignment.
	5e. Scum baffle not at correct depth.	5e. Baffle for scum by-passing.	5e. Increase baffle depth.

Indicators/Observations	Possible Cause	Check or Monitor	Possible Solutions
6. Foaming.	6a. A foaming substance has entered the plant.	6a. Industrial contributors.	6a1. Identify and restrict this material from entering the plant if it interferes with plant operation. 6a2. Use spray water or foam reducing chemicals to keep foam down.
7. Sludge settling in influent lines.	7a. Velocity too low.	7a. Velocity.	7a1. Use fewer tanks to increase velocity. 7a2. Agitate with air or water to keep the material in suspension until it reaches the tank.

PRIMARY CLARIFICATION

Indicators/Observations	Possible Cause	Check or Monitor	Possible Solutions
8. Floating sludge.	8a. Sludge becoming septic in tank.	8a. Tank surface for black sludge.	8a. Remove sludge more frequently or at a higher rate.
	8b. Damaged or worn collection equipment.	8b. Inspect collection equipment.	8b. Repair or replace as necessary.
	8c. Recycled waste sludge.	8c. Quantity and quality of recycled waste sludge.	8c. Recycle at periods of low flow or recycle to different treatment process.
	8d. Primary sludge pump malfunction.	8d. Pump performance.	8d. Troubleshoot pump operation. Place a standby pump in operation.
	8e. Sludge withdrawal line plugged.	8e. Sludge pump output.	8e. Clean line.
	8f. Return of well nitrified waste activated sludge.	8f. Effluent nitrates.	8f. Vary the age of returned sludge, or recycle waste sludge to an alternate process.
	8g. Too few tanks in service.	8g. Retention time.	8g. Place more tanks into service.
	8h. Damaged or missing baffles.	8h. Baffles.	8h. Repair, replace, or install baffles.

# WASTEWATER TREATMENT TROUBLESHOOTING AND PROBLEM SOLVING

Indicators/Observations	Possible Cause	Check or Monitor	Possible Solutions
9. Septic wastewater or sludge.	9a. Damaged or worn collection equipment.	9a. Sludge collectors.	9a. Repair or replace as necessary.
	9b. Infrequent sludge removal.	9b. Sludge density.	9b. Increase frequency and duration of sludge removal until sludge density is correct for your plant.
	9c. Insufficient industrial pretreatment.	9c. Pretreatment.	9c1. Improve industrial pretreatment standards.
			9c2. Aerate tank influent.
	9d. Septic sewage from collection system.	9d. Retention time and velocity in collection system.	9d1. Reduce retention time or increase velocity in collection system.
			9d2. Add an oxidant (usually chlorine) to collection system.
			9d3. Aerate wastewater.
	9e. Strong recycle flows (usually digester supernatant).	9e. Recycle quality and quantity.	9e1. Improve digester operation to produce a milder supernatant.

PRIMARY CLARIFICATION

Indicators/Observations	Possible Cause	Check or Monitor	Possible Solutions
			9e2. Provide treatment before recycle.
			9e3. Reduce rate of return.
			9e4. Recycle at periods of low flow.
			9e5. Recycle to a different treatment process
	9f. Primary sludge pump malfunction.	9f. Pump performance.	9f. Troubleshoot pump operation. Place a standby pump into operation.
	9g. Sludge withdrawal line plugged.	9g. Sludge pump output.	9g. Clean line.
	9h. Sludge collectors not run often enough.	9h. Sludge density.	9h. Increase run time or run continuously.
	9i. Septage dumpers.	9i. Sample trucks.	9i. Monitor, regulate, or restrict septage truck dumping.

Indicators/Observations	Possible Cause	Check or Monitor	Possible Solutions
10. Primary sludge solids concentrations are too low.	10a. Hydraulic overload.	10a. Influent flow rate.	10a. Provide more tanks or more even flow through tanks.
	10b. Overpumping of sludge.	10b. Frequency and duration of sludge pumping. Suspended solids concentration.	10b. Reduce frequency and duration of pumping cycles. Check blanket levels.
	10c. Collection system problem.	10c. Collection system.	10c. Troubleshoot collection system.
	10d. Decreased influent solids loading.	10d. Influent suspended solids loading.	10d. Take one or more primary tanks out of service.

PRIMARY CLARIFICATION

Indicators/Observations	Possible Cause	Check or Monitor	Possible Solutions
11. Primary sludge solids concentrations are too high. Hard to remove from hopper.	11a. Excessive grit and compacted material.	11a. Grit removal system.	11a. Improve operation of grit removal equipment.
	11b. Primary sludge pump malfunction.	11b. Pump performance.	11b. Troubleshoot pump operation. Place a standby pump into operation.
	11c. Sludge withdrawal line plugged.	11c. Sludge flow.	11c. Clean line.
	11d. Sludge retention time is too long.	11d. Blanket levels.	11d. Increase operation of collection equipment and primary sludge pumping rate.
	11e. Increased influent loadings.	11e. Influent suspended solids loading.	11e. Check blanket levels. Increase operation of collection equipment and primary sludge pumping rate.

Indicators/Observations	Possible Cause	Check or Monitor	Possible Solutions
12. Excessive growth or debris on surfaces and weirs.	12a. Accumulation of solids and growths.	12a. Inspect surfaces.	12a. Clean surfaces.
	12b. Poor housekeeping.	12b. Inspect surfaces.	12b. Clean more frequently.
13. Collector stops or binds.	13. Physical jam.	13. Check for an obstruction, high sludge level, or scraper plow scraping floor.	13a. If sludge level is high enough to bind collectors, determine cause and correct. Increase sludge removal rates.
			13b. Drain tank and check mechanism for free operation. Remove object.
			13c. Adjust scraper plow up from floor.

PRIMARY CLARIFICATION   37

Indicators/Observations	Possible Cause	Check or Monitor	Possible Solutions
14. Erratic operation of collectors.	14a. Drive chain has too much slack.	14a. Chain tension.	14a. Adjust chain slack (remove one or more chain links).
	14b. Sludge build-up.	14b. Sludge levels.	14b. Check sludge levels and increase sludge pumping if necessary.
	14c. Rags and debris have tangled collector.	14c. Sludge collector.	14c. Remove tangles.
	14d. Damaged collectors or broken shear pins.	14d. Collector and shear pins.	14d. Repair or replace damaged parts.
15. Collector will not start.	15a. Breaker tripped.	15a. Circuit breaker.	15a. Reset breaker and attempt to restart.
	15b. Motor failure.	15b. Test motor.	15b. Replace motor.
16. Skimmers not skimming properly.	16a. Flight or blade height not properly set.	16a. Flight or blade height.	16a. Adjust to proper setting.
	16b. Beach not properly adjusted (circular tank).	16b. Beach and skimmer contact.	16b. Adjust height.

NOTES

7 LAGOONS/PONDS

7.1. PROCESS OVERVIEW

- Lagoons, or stabilization ponds, treat wastewater through the use of sunlight, wind, algae, and oxygen.

- In stabilization ponds, wastewater enters the pond at a single point, either in the middle or at the edge. Algae grow in the pond by taking energy from the sunlight and using up the carbon dioxide and inorganics released by bacteria. The algae releases oxygen for use by the bacteria.

- Lagoons are usually deeper than the ponds. An unaerated or facultative lagoon is usually 3 to 5 feet deep and uses aerobic and facultative bacteria to break down the organic material in the wastewater. Oxygen is furnished by algae and wind action. A facultative pond generally has an aerobic zone near the surface, an aerobic and anaerobic zone in the middle depth, and an anaerobic zone at the bottom.

- In an aerated lagoon, aeration equipment is installed to provide additional oxygen. Air may be supplied by a compressor through tubes installed in the lagoon bottom or by mechanical aerators installed at the surface. Aerated ponds are deeper (10 to 18 feet) and smaller than facultative or unaerated lagoons.

- Key factors in lagoon/pond operation include the following concepts:

    $$- \text{Area, acres} = \frac{(\text{width, feet})(\text{length, feet})}{43,560 \text{ square feet per acre}}$$

- Volume, acre-feet = (area, acres)(depth, feet)

- Flow rate, $\dfrac{\text{acre-feet}}{\text{day}} = \dfrac{\text{Flow into pond, gallons per day}}{(7.48 \text{ gal/cu ft})(43{,}560 \text{ sq ft/acre})}$

- Detention time, days $= \dfrac{(\text{Volume, acre-feet})}{\text{flow, acre-feet/day}}$

- Organic loading rate, $\dfrac{\text{lbs/day}}{\text{acre}} = \dfrac{\text{BOD into pond, lbs/day}}{\text{pond area, acres}}$

7.2. LAGOON/POND TROUBLESHOOTING

Indicators/Observations	Possible Cause	Check or Monitor	Possible Solutions
1. Excessive weed growth.	1. Poor circulation or maintenance. Shallow water.	1. Circulation patterns. Water depth.	1a. Pull weeds by hand. 1b. Mow weeds. 1c. Lower water level and burn weeds with a torch. 1d. Lower water level. Raise level back after water freezes to pull out weeds. 1e. Increase water depth to drown weeds. 1f. For duckweed, use a boat and push the duckweed with a board to an area where it can be physically removed. 1g. Use an approved herbicide.

Indicators/Observations	Possible Cause	Check or Monitor	Possible Solutions
			1h. Install a pond liner.
			1i. Be sure water depth is at least 3 feet.
2. Burrowing animals.	2. Bank conditions may attract local animals.	2. Observe for animal activity.	2a. Remove food source (cattails and reeds).
			2b. Raise and lower the water level for a week at a time to alternately drown or expose holes.
3. Scum formation.	3. Pond bottom turning over with sludge floating to top. Poor circulation. Grease and oil in influent.	3. Observe pond/lagoon service.	3a. Break up scum using rakes, water jets, or a boat. (Broken scum will settle).
			3b. Skim remaining scum for approved disposal.
4. Blue-green algae growth.	4. Algae growth indicates poor treatment, overloading, or poor nutrient balance.	4. Pond surface.	4a. Apply copper sulfate (5 lbs/million gallons).
			4b. Break up algae manually.

LAGOONS/PONDS 43

Indicators/Observations	Possible Cause	Check or Monitor	Possible Solutions
5. Odors.	5a. Overloading, poor circulation, industrial wastes, and anaerobic conditions.	5a. Dissolved oxygen (D.O.), total and dissolved sulfides.	5a1. Break up and resuspend septic sludge and scum. 5a2. If available, use extra cell. 5a3. Add sodium nitrate. 5a4. Prechlorinate influent flow. 5a5. Recirculate the effluent (1:6 ratio). 5a6. Install floating aerators. 5a7. Eliminate septic or industrial wastes.
	5b. Algae growths.	5b. Check surface for algae.	5b. See 4a and 4b above.

Indicators/Observations	Possible Cause	Check or Monitor	Possible Solutions
6. Insects.	6. Poor circulation and maintenance.	6. Visual inspection.	6a. Remove weeds and scum to encourage wave action.
			6b. Stock pond with fish.
			6c. Apply insecticide as a last resort. (Check with regulatory agency).
7. Inability to maintain sufficient liquid level.	7a. Leakage or percolation.	7a. Look for seepage around dikes.	7a1. Apply bentonite clay to seal leak.
			7a2. Install a liner.
	7b. Evaporation.	7b. Check detention time. May be too long.	7b. Divert extra water, land drainage or stream flow, into lagoon.

LAGOONS/PONDS 45

Indicators/Observations	Possible Cause		Check or Monitor		Possible Solutions	
8. Groundwater contamination.	8.	Leakage.	8.	Seepage. Use monitoring wells if available.	8.	Apply bentonite clay or install liner to seal leaks.
9. Low dissolved oxygen (should be at least 3.0 in warm weather).	9a.	Poor light penetration and low algae growth.	9a.	Pond color may be gray.	9a.	Remove weeds and floating debris.
	9b.	Low detention time.	9b.	Check detention time.	9b.	Increase detention time.
	9c.	Poor circulation.	9c.	D.O.	9c.	Add floating aerators or sodium nitrate.
	9d.	Hydrogen sulfide in influent.	9d.	Hydrogen sulfide odor.	9d.	Eliminate septic influent or chlorinate influent.
	9e.	High BOD loading or toxic/industrial wastes	9e.	Influent characteristics.	9e.	Eliminate high strength source or go to parallel operation.
	9f.	Poor wind action.	9f.	Check tree growth.	9f.	Cut trees and growth near pond to allow wind to get to pond.

Indicators/Observations	Possible Cause	Check or Monitor	Possible Solutions
10. Incomplete waste treatment.	10a. Overloading due to short circuiting, industrial wastes, undersized lagoon, or addition of a new service area.	10a. Look for a yellow, green or gray color, a low pH and D.O., and high BOD.	10a1. Use parallel operation. 10a2. Recirculate pond effluent. 10a3. Look for and correct a possible short-circuit. 10a4. Add aeration.
11. Decreasing pH.	11a. Overloading, poor weather, or loss of algae.	11a. Check pH (should be 8.0 to 8.4). Check D.O.	11a1. Look for and eliminate a possible toxic contributor. 11a2. Use parallel operation. 11a3. Recirculate pond effluent. 11a4. Look for and correct a possible short circuit. 11a5. Add aeration.

LAGOONS/PONDS 47

Indicators/Observations	Possible Cause	Check or Monitor	Possible Solutions
12. Facultative pond has turned anaerobic.	12a. Overloading, a toxic discharge, or short circuiting.	12a. Check BOD, suspended solids, and look for scum, odors, and a yellow or green color.	12a1. Change from series to parallel operation to divide load. 12a2. Change inlets and outlets to stop short circuiting. 12a3. Add aerators and/or recirculate effluent.

48 WASTEWATER TREATMENT TROUBLESHOOTING AND PROBLEM SOLVING

Indicators/Observations	Possible Cause	Check or Monitor	Possible Solutions
13. Poor effluent quality.	13a. Organic overloading.	13a. Check BOD.	13a. Add aerators and/or recirculate effluent.
	13b. Low temperature.	13b. Check air temperature. Look for brown color in pond.	13b. Operate ponds in series.
	13c. Toxic influent.	13c. Brown color in pond.	13c. Identify and eliminate source.
	13d. Loss of lagoon volume due to accumulated sludge.	13d. Check pond depth (sludge depth).	13d. Remove sludge more frequently.
	13e. Aeration equipment failure.	13e. Check aeration equipment.	13e. Repair aeration equipment.
	13f. Excess turbidity from scum or algae mats.	13f. Check effluent turbidity.	13f. Break up mats.
	13g. Light blocked by excessive plant growth.	13g. Inspect plant growth.	13g. Cut plant growth at regular intervals.

NOTE: Do not apply an herbicide or insecticide if its application could cause a toxicity violation. Discuss any possible use of an herbicide or insecticide with your local regulatory officials before application.

NOTES

8 TRICKLING FILTERS

8.1. PROCESS OVERVIEW

- A trickling filter consists of a bed of coarse media, usually rocks or plastic, covered with microorganisms. The wastewater is applied to the media at a controlled rate, using a rotating distributor arm or fixed nozzles. Organic material is removed by contact with the microorganisms as the wastewater trickles down through the media openings. The treated wastewater is then collected by an underdrain system.

- The trickling filter is usually built into a tank which contains the media. The filter may be square, rectangular, or circular.

- The trickling filter does not provide any actual filtration, the filter media provides a large amount of surface area that the microorganisms can cling to and grow in a slime that forms on the media as they feed on the organic material in the wastewater.

- The slime growth on the trickling filter media periodically sloughs off and is settled and removed in a secondary clarifier which follows the filter.

- Key factors in trickling filter operation include the following concepts:

    - Hydraulic loading rate,

    $$\frac{\text{gal/day}}{\text{sq/ft}} = \frac{\text{Flow, gal/day (including recirc.)}}{\text{Media top surface, sq/ft}}$$

- Organic loading rate,

$$\frac{\text{lbs/day}}{1{,}000 \text{ cu/ft}} = \frac{\text{BOD into filter, pounds per day}}{\text{Media volume, } 1{,}000 \text{ of cu/ft}}$$

- Recirculation, ratio $= \dfrac{\text{Recirculation flow, MGD}}{\text{Average influent flow, MGD}}$

8.2. TRICKLING FILTER TROUBLESHOOTING

Indicators/Observations	Possible Cause	Check or Monitor	Possible Solutions
1. Filter ponding.	1a. Media too small or not uniform in size.	1a. Check size of media.	1a. Replace media.
	1b. Rock media broken (freeze fracture damage).	1b. Broken pieces clogging media.	1b. Replace media.
	1c. Poor primary treatment.	1c. Excessive suspended solids (SS) in filter influent.	1c. Correct primary treatment problems.
	1d. Excessive biological growth and sloughing.	1d. Slime growths clogging filter.	1d. Break up growth with a high pressure stream of water, apply chlorine, or shut down filter to dry out media growths.
	1e. Excessive organic loading.	1e. Check loading rate.	1e. Increase recirculation or flood media to loosen and remove accumulated growth.
	1f. Accumulation of trash on media.	1f. Inspect filter surface.	1f. Clean trash from filter.
	1g. Snails, moss or roaches.	1g. Inspect media.	1g. Flush media or chlorinate.

54 WASTEWATER TREATMENT TROUBLESHOOTING AND PROBLEM SOLVING

Indicators/Observations	Possible Cause	Check or Monitor	Possible Solutions
2. Filter flies.	2a. Excessive biological growths.	2a. Inspect media.	2a. Remove excessive growths as described above.
	2b. Unkept grounds.	2b. Inspect grounds.	2b. Maintain grounds (cut grass, housekeeping chores, etc.).
	2c. Low hydraulic loading.	2c. Increase hydraulic loading to at least 200 gpd/sq ft.	2c. Disrupt fly life cycle by increasing recirculation, flooding the filter, chlorinating to a 1.0 chlorine residual, and applying an insecticide to walls and breeding areas.
	2d. Poor wastewater distribution.	2d. Inspect distribution pattern.	2d. Unclog spray nozzles or increase flow.
Note: Do not apply insecticide if its application could cause a toxicity violation. Discuss any possible use of insecticide with your local regulatory officials before application.			

TRICKLING FILTERS

Indicators/Observations	Possible Cause	Check or Monitor	Possible Solutions
3. Odors.	3a. Excessive organic loading.	3a. Check organic loading.	3a1. Maintain aerobic conditions (use forced air if necessary).
			3a2. Chlorinate filter influent when flow is low.
			3a3. Increase recirculation rate or add oxidant.
	3b. Poor ventilation due to clogged vent or drain.	3b. Check vent and drain.	3b. Clear vent and drain system. Reduce hydraulic loading if underdrains are flooded.
	3c. Poor ventilation due to excessive biological growth.	3c. Inspect media.	3c. Increase recirculation rate.
	3d. Trash on media.	3d. Inspect media.	3d. Remove trash.
	3e. Septic influent or industrial discharge.	3e. Check influent for H_2S or industrial wastes.	3e. Correct or remove cause. Aerate or prechlorinate.

56 WASTEWATER TREATMENT TROUBLESHOOTING AND PROBLEM SOLVING

Indicators/Observations	Possible Cause	Check or Monitor	Possible Solutions
4. Ice buildup or media	4a. Cold weather.	4a. Air and wastewater temperature.	4a1. Decrease recirculation.
			4a2. Operate two-stage systems in parallel.
			4a3. Adjust distributors for coarser spray.
			4a4. Partially open end dump gates on distributors.
			4a5. Break up and remove ice.
			4a6. Reduce number of filters in service.
			4a7. Reduce detention time in preliminary and primary treatment units.
	4b. Uneven wastewater distribution.	4b. Inspect distribution.	4b. Adjust distributors to provide a more even flow.

TRICKLING FILTERS 57

Indicators/Observations	Possible Cause	Check or Monitor	Possible Solutions
5. Uneven flow distribution.	5a. Clogged distributor orifices.	5a. Ponds on some surface area, dry spots on others.	5a. Remove, clean, and flush distributor openings.
	5b. Low hydraulic load.	5b. Hydraulic loading rate.	5b. Recirculate or take other actions to maintain hydraulic load.
	5c. Leaking seal.	5c. Check seal.	5c. Replace seal.
6. Snails, moss, and roaches.	6. Warm climate.	6. Inspect filter surface.	6. Flush filter with maximum recirculation and chlorinate to a 1.0 mg/L residual.

58   WASTEWATER TREATMENT TROUBLESHOOTING AND PROBLEM SOLVING

Indicators/Observations	Possible Cause	Check or Monitor	Possible Solutions
7. Increased clarifier effluent suspended solids and/or BOD.	7a. Excess filter sloughing due to seasonal change.	7a. Seasonal effect on microorganism.	7a. Add polymer to clarifier influent.
	7b. Excessive sloughing due to heavy organic loading.	7b. Check organic loading.	7b1. Increase clarifier underflow rate.
			7b2. Eliminate return of any high strength side streams
	7c. Excessive sloughing due to pH or toxic conditions.	7c. Check for toxic chemicals or pH change.	7c. Maintain pH between 6.5 and 8.5. Identify and eliminate source of toxic wastewater causing upset.
	7d. Denitrification in clarifier.	7d. Check for nitrification in effluent and look for floating sludge or bubbles.	7d1. Increase clarifier underflow rate.
			7d2. Increase loading on the tricking filter and remove or break up any sludge clumps.

TRICKLING FILTERS

Indicators/Observations	Possible Cause	Check or Monitor	Possible Solutions
	7e. Clarifier hydraulically overloaded.	7e. Check overflow rate (should not exceed 1,200 gpd/sq ft.).	7e. Reduce recirculation rate during peak flow periods.
	7f. Equipment malfunction in clarifier.	7f. Check for broken or failed equipment.	7f. Replace or repair any broken or failed equipment.
	7g. Short-circuiting of flow through clarifier.	7g. Check weirs or add a dye to trace flow.	7g. Level effluent weirs and/or install baffles in the clarifier.
	7h. Undesirable growth on filter.	7h. Observe growth under a microscope.	7h. Chlorinate filter to kill off undesirable growth.

60 WASTEWATER TREATMENT TROUBLESHOOTING AND PROBLEM SOLVING

Indicators/Observations		Possible Cause		Check or Monitor		Possible Solutions
8. Rotating arm slows or stops.	8a.	Insufficient flow.	8a.	Check hydraulic loading.	8a.	Increase hydraulic loading.
	8b.	Clogged orifices.	8b.	Check arm for trash.	8b.	Open dump ends and remove trash. Prevent solids from reaching the filter.
	8c.	Clogged arm vent.	8c.	Check vent.	8c.	Rod or flush vent pipe to remove solids.
	8d.	Bad main bearing.	8d.	Check bearing.	8d.	Replace bearing if necessary.
	8e.	Support wires loose (arm not level).	8e.	Check wire tension.	8e.	Tighten at tie rods.
	8f.	Distributor arm hitting media.	8f.	Check media level.	8f.	Remove some media from high spots.
9. Dirt in main bearing lube oil.	9a.	Worn seals.	9a.	Check seals.	9a.	Replace if necessary.
	9b.	Oil level too low.	9b.	Check oil level.	9b.	Refill oil if necessary.
10. Water leak at distributor base.	10a.	Worn seal.	10a.	Check seal.	10a.	Replace seal if necessary.
	10b.	Worn expansion joint.	10b.	Check expansion joint.	10b.	Repair or replace expansion joint.

TRICKLING FILTERS 61

Indicators/Observations	Possible Cause	Check or Monitor	Possible Solutions
11. Secondary clarifier sludge collector stopped.	11a. Torque overload.	11a. Check blanket level and for objects in tank.	11a. Lower blanket level or remove foreign objects.
	11b. Loss of power.	11b. Check breakers.	11b. Reset breaker. If collector stops again, have maintenance determine cause.
	11c. Drive unit failure.	11c. Check drive chain and shear pins.	11c. Repair or replace broken equipment and restart.
12. Recirculation pump delivering insufficient flow.	12a. Excessive head.	12a. Check for a closed valve or clogged pipes.	12a. Open valves and clear obstructions.
	12b. Pump malfunction.	12b. Check packing, bearings, and pump casing for air lock.	12b. Replace packing, lubricate bearings, or bleed air out of pump casing where appropriate.
	12c. Pump motor failure.	12c. Check breaker, bearings, alignment, and current draw.	12c. Reset breaker, lubricate bearings, realign pump, or have motor rewound where appropriate.

NOTES

9 ROTATING BIOLOGICAL CONTRACTORS (RBCs)

9.1. PROCESS OVERVIEW

- A Rotating Biological Contractor (RBC) uses a biological slime of microoganisms that grow on thin disks made of plastic mounted side by side on a shaft.

- The shaft mounted disks are rotated slowly so that they are alternately submerged in the incoming wastewater then exposed to air. (The speed can be adjusted.)

- The microbes in the wastewater begin to stick to the disk surfaces and grow there until all the disks are covered with a thin layer of biological slime. The microbes get oxygen from the air as the disks are rotated.

- The excess growth breaks off from the disks and flows out to a secondary clarifier where it is separated from the wastewater by sedimentation.

- The attached growth is similar to that of the trickling filter except here the microbes are rotated into the wastewater rather than the wastewater being distributed over the microbes.

9.2. ROTATING BIOLOGICAL CONTRACTORS TROUBLESHOOTING

Indicators/Observations	Possible Cause	Check or Monitor	Possible Solutions
1. Excessive sloughing of biomass.	1a. Toxic material in influent.	1a. Identify material and its source.	1a. Eliminate toxic source or use flow equalization to dampen its effect.
	1b. pH variations.	1b. Check pH.	1b. Eliminate unusual pH source or adjust pH to within the operable range of between 5 and 10.
2. Development of a white biomass.	2a. Septic influent (high H_2S).	2a. Check for H_2S level or odor.	2a. Preaerate wastewater, add oxygen under discs, or add sodium nitrate or hydrogen peroxide to boost oxygen.
	2b. First stage is organically overloaded.	2b. Check organic loading on first stage.	2b. Adjust baffles between first and second stages to increase first stage total surface area. Add additional treatment units.

Indicators/Observations	Possible Cause	Check or Monitor	Possible Solutions
3. Decreased treatment efficiency.	3a. Organic overload.	3a. Check peak organic load. Should not be the cause if this is less than twice the daily average.	3a. Improve pretreatment. Add treatment units.
	3b. Hydraulic overload.	3b. Check peak hydraulic load. Should not be the cause if this is less than twice the daily average.	3b. Equalize flow between reactors or eliminate source of excess flow.
	3c. pH too high or too low.	3c. Check influent pH.	3c. Eliminate unacceptable pH source or adjust pH. Best pH range is 6.5 to 8.5 for secondary treatment, 8.0 to 8.5 for nitrification. For nitrification, be sure alkalinity is 7 times the influent ammonia concentration.

Indicators/Observations	Possible Cause	Check or Monitor	Possible Solutions
	3d. Low wastewater temperature.	3d. Check temperature.	3d. Low (<55°F) temperature reduces treatment efficiency. Add treatment units if available.
	3e. Too many snails.	3e. Evaluate treatment performance.	3e. Periodically remove unit from service and clean media with a caustic solution.
4. Solids buildup in reactor troughs.	4. Inadequate pretreatment.	4. Determine if solids are organic or grit.	4. Remove solids and improve grit removal or primary treatment.
5. Snails.	5a. Environment conducive to snail growth.	5a. Check for snail accumulation.	5a. Periodically remove unit from process and clean with a caustic or chlorine solution and/or increase RBC speed.
	5b. Low organic loading.	5b. Check organic load.	5b. Rearrange loading patterns to increase RBC load.

ROTATING BIOLOGICAL CONTRACTORS (RBCs) 67

Indicators/Observations	Possible Cause	Check or Monitor	Possible Solutions
6. Shaft bearing failure.	6. Inadequate preventive maintenance.	6. Check maintenance log.	6. Lubricate bearings. Replace if necessary.
7. Motors overheating.	7a. Inadequate maintenance.	7a. Check oil levels.	7a. Lubricate.
	7b. Improve chain drive alignment.	7b. Check alignment.	7b. Correct alignment.

NOTES

10 ACTIVATED SLUDGE

10.1. PROCESS OVERVIEW

- The activated sludge process is a treatment technique in which wastewater and reused biological sludge full of living microorganisms is mixed and aerated. The biological solids are then separated from the treated wastewater in a clarifier and returned to the aeration process or wasted.

- The microorganisms are mixed thoroughly with the incoming organic material and they grow and reproduce by using the organic material as food. As they grow and are mixed with air, the individual organisms clump together (flocculate). Once flocculated, they more readily settle in the secondary clarifier.

- The wastewater being treated flows continuously into an aeration tank where air is injected to mix the wastewater with the returned activated sludge and to supply the oxygen needed by the microbes to live and feed on the organics. Aeration can be supplied by injection through air diffusers in the bottom of the tank or by mechanical aerators located at the surface.

- The mixture of activated sludge and wastewater in the aeration tank is called the "mixed liquor." The mixed liquor flows to a secondary clarifier where the activated sludge is allowed to settle.

- The activated sludge is constantly growing and more is produced than can be returned for use in the aeration basin. Some of this sludge must, therefore, be wasted to a sludge handling system for treatment and disposal. The volume of sludge returned to the aeration basins is normally 40 to 60 percent of the wastewater flow. The rest is wasted.

- Key factors in activated sludge operation include the following concepts:

- Sludge Volume Index (SVI), mL/g =

$$\frac{(30 \text{ min settled volume, mL/L}) (1000)}{\text{mixed liquor suspended solids, mg/L}}$$

- Food to Microorganism (F/M), ratio =

$$\frac{(\text{BOD or COD in primary effluent, lbs/day})}{\text{pounds volatile SS in aeration tanks}}$$

- Mean Cell Residence Time (MCRT), days =

$$\frac{\text{solids in total system, lbs}}{\text{solids wasted + lost, lbs/day}}$$

Note: MCRT may also be referred to as Sludge Retention Time (SRT).

10.2. ACTIVATED BIOFILTER (ABF) PROCESS OVERVIEW

Note: ABF is covered here with activated sludge because although the ABF treatment is similar to both trickling filters and activated sludge, its problems and responses are more similar to the activated sludge process.

- A fixed growth of microorganisms develops on redwood slats or synthetic media similar to trickling filters or RBC units except here settled sludge from the secondary clarifier is recycled over the ABF.

- By means of sludge recirculation, a population of suspended growth microbes is developed in addition to the fixed growth on the racks or media.

- The ABF is also deeper than a trickling filter; usually above 14 feet of media or redwood is used.

- Oxygen is supplied by the splashing of the wastewater onto the top of the ABF by a distributor and by the movement of the wastewater over the microbes and down through the ABF.

- A completely mixed aeration basin sometimes follows the ABF to help oxidize organics and improve settling. Part of the ABF underflow passes to the aeration basin and the rest is returned to the wet well for recirculation.

- A secondary clarifier is used for final sedimentation with the sludge recirculated to the ABF.

10.3. ACTIVATED SLUDGE TROUBLESHOOTING

Indicators/Observations	Possible Cause	Check or Monitor	Possible Solutions
1. Rising sludge in secondary clarifier.	1a. Growth of filamentous organisms in activated sludge ("bulking sludge").	1a. Sludge Volume Index (SVI). If less than 100, filaments are not cause. View sludge under microscope to identify filaments.	1a1. Increase D.O. in aeration tank if less than 1.0 mg/L. 1a2. Increase pH to 7.0. 1a3. Check for nutrient deficiency. BOD to nutrient ratio should be no more than 100 mg/L BOD to 5 mg/L total nitrogen, to 1 mg/L phosphorous, and 0.5 mg/L iron. Add nutrient sources if needed. 1a4. Add up to 50 mg/L chlorine or up to 200 mg/L hydrogen peroxide to return sludge until SVI drops below 150. 1a5. Increase mean cell residence time (MCRT). 1a6. Increase sludge return rate.

72   WASTEWATER TREATMENT TROUBLESHOOTING AND PROBLEM SOLVING

Indicators/Observations	Possible Cause		Check or Monitor		Possible Solutions	
					1a7.	Add a settling aid such as polymer.
	1b.	Denitrification.	1b.	Check for nitrates in effluent. Denitrification is likely if nitrates exceed 10 mg/L. Check pH of mixed liquor (decreases if denitrification is occurring). Check chlorine demand (an increase in demand indicates denitrification).	1b.	Decrease aeration tank detention by taking an aeration tank out-of-service to decrease nitrification, or reduce mixed liquor suspended solids (MLSS) by increasing wasting (WAS) rates. This increases organic loading and cuts down nitrification.
	1c.	Overaeration.	1c.	Check D.O.	1c.	Decrease aeration if lowest D.O. reading in aeration basin is more than 4.0.
	1d.	Septic sludge due to not removing sludge fast enough.	1d.	Check for sulfides.	1d.	Increase return (RAS) rates to shorten time sludge is in clarifier and/or add an oxidizing agent such as chlorine or hydrogen peroxide.

ACTIVATED SLUDGE 73

Indicators/Observations	Possible Cause	Check or Monitor	Possible Solutions
2. Pin floc in secondary clarifier effluent.	2a. Excessive aeration tank mixing.	2a. Aeration tank appearance.	2a. Reduce aeration agitation.
	2b. Overaeration.	2b. D.O. in aeration tank. (Should not be over 4.0).	2b. Reduce aeration and increase sludge wasting to bring MCRT down.
	2c. Anaerobic conditions in aeration tank.	2c. D.O. in aeration tank (Should be at least 2.0).	2c. Increase D.O. in aeration tank.
	2d. Toxic shock load (possible industrial bypass).	2d. Look for dead protozoans in the sludge with a microscope.	2d. Identify and eliminate source of toxic load. Reseed with activated sludge from another plant or add freeze-dried microorganisms.
3. Dark, stiff tan foam on aeration basin.	3. MCRT is too long.	3. Check MCRT. Should not be longer than 9 days.	3. Increase sludge wasting to reduce MCRT.

74 WASTEWATER TREATMENT TROUBLESHOOTING AND PROBLEM SOLVING

Indicators/Observations	Possible Cause	Check or Monitor	Possible Solutions
4. White, sudsy foam on aeration basin.	4a. MLSS is too low (excess wasting).	4a. Check MLSS.	4a. Decrease sludge wasting to increase MLSS.
	4b. Could be any or a combination of the following: toxic waste, nutrient deficiency, too high or low pH, or insufficient D.O.	4b. Monitor plant influent. Check for a low respiration rate. Test mixed liquor for metals or nutrient deficiency.	4b. Waste sludge and seed with new sludge from another plant or use the freeze-dried variety. Identify and correct problem caused by toxic or problem influent.
	4c. Aeration tank overload.	4c. Check influent strength and wasting rates.	4c. Reduce wasting/increase MLSS to compensate for heavier loading.

ACTIVATED SLUDGE 75

Indicators/Observations	Possible Cause	Check or Monitor	Possible Solutions
5. Sludge blanket lost out of clarifier. Aeration basins look dark.	5. Insufficient aeration.	5. D.O. in aeration basin (should be at least 2.0).	5a1. Increase aeration. 5a2. Decrease loading (put another aeration basin on line). 5a3. Check aeration piping for leaks or closed valves. 5a4. Clean plugged diffusers.
6. MLSS concentrations different in each aeration basin.	6. Unequal flow distribution to the aeration basins.	6. Check flow to each aeration basin.	6. Adjust flow control devices to equalize flow to each aeration basin.

Indicators/Observations	Possible Cause	Check or Monitor	Possible Solutions
7. Sludge blanket uniformly overflowing weirs in secondary clarifiers.	7a. Inadequate sludge return rate.	7a. Check return sludge pump flow.	7a1. Put another pump on line if pump has malfunctioned.
			7a2. Increase sludge pump return flow rate and monitor sludge blanket. Maintain a 1- to 3-foot blanket.
			7a3. Clean sludge return line if line is plugged.
	7b. Hydraulic overload to a clarifier due to unequal flow distribution.	7b. Check flow to each clarifier.	7b. Adjust flow control devices to equalize flow to each clarifier.
	7c. Peak flows are overloading clarifiers.	7c. Check peak flow overflow rates. Should not exceed 1,000 gpd/sf.	7c. Use all available clarifiers or divert flow to another process.
	7d. Solids loading rate is too high.	7d. Check solids loading rate. Solids loading should not exceed 1.25 lb/sf/hr.	7d. Reduce MLSS concentration by increasing wasting rates.

ACTIVATED SLUDGE 77

Indicators/Observations	Possible Cause	Check or Monitor	Possible Solutions
8. Sludge blanket overflowing weirs in one area of secondary clarifier.	8. Unequal flow distribution within clarifier.	8. Check effluent weir level.	8. Level effluent weirs.
9. Air boiling or rising in large bubbles in aeration basin.	9. Plugged or broken diffusers.	9. Visually inspect aeration diffusers.	9. Clean or replace diffusers.
10. Dead spots in aeration basin.	10a. Plugged diffusers.	10a. Visually inspect aeration diffusers.	10a. Clean or replace diffusers.
	10b. Low D.O. due to insufficient aeration.	10b. Check D.O.	10b. Increase aeration to assure at least a 2.0-mg/L D.O. content.
11. Sludge becomes less dense, mixed liquor pH decreases to 6.7 or lower.	11a. Nitrification. Wastewater alkalinity is low.	11a. Check effluent ammonia level and alkalinity of influent and effluent.	11a1. Decrease sludge age by increased wasting if nitrification not desired.
			11a2. Add sodium bicarbonate or another source of additional alkalinity.
	11b. Acidic influent.	11b. Check influent pH.	11b. Identify and eliminate source of acidic wastewater.

Indicators/Observations	Possible Cause	Check or Monitor	Possible Solutions
12. Return sludge concentration is too thin (less than 8,000 mg/L).	12a. Sludge return rate is too high.	12a. Check return sludge concentration. Perform settleability test.	12a. Reduce sludge return rate.
	12b. Growth of filamentous organisms.	12b. Check D.O., pH, and nitrogen concentration. Inspect return sludge sample under a microscope.	12b. Increase D.O. and pH, add nitrogen source, and add chlorine.
	12c. Presence of Actinomycetes bacteria.	12c. Check dissolved iron (ferrous) content. Inspect return sludge sample under a microscope.	12c. Add an iron source if the dissolved iron content is less than 0.5 mg/L.

NOTES

11 SECONDARY CLARIFIERS

11.1. PROCESS OVERVIEW

- Secondary clarifiers are constructed and operated like the primary clarifiers except that they follow a biological treatment process such as a trickling filter or activated sludge. The type of biological treatment they follow dictates how they are operated.

- Secondary clarifiers following trickling filters or RBCs are used to settle biological solids that have separated from the media.

- Secondary clarifiers following activated sludge processes help settle and clarify the effluent while also providing a concentrated source of return sludge for process control.

- Secondary clarifiers may be round or rectangular, and may be designed for natural gravity settling or enhanced chemically aided settling.

- Key factors in secondary clarifier operation include the following concepts:

  - Retention time, hours = $\dfrac{(\text{volume, gal})\,(24\text{ hrs/day})}{\text{flow, gallons per day}}$ (inc RAS)

  - Surface loading rate, $\dfrac{\text{gal/day}}{\text{sq/ft}} = \dfrac{\text{flow, gallons per day}}{\text{surface area, sq/ft}}$ (w/o RAS)

  - Surface loading rate, $\dfrac{\text{lbs/day}}{\text{sq/ft}} = \dfrac{\text{solids into clarifier, lbs/day}}{\text{surface area, sq/ft}}$ (inc RAS)

- Weir overflow rate, $\dfrac{\text{gal/day}}{\text{lineal ft}} = \dfrac{\text{flow, gallons per day}}{\text{weir length, lineal ft}}$ (w/o RAS)

11.2. SECONDARY CLARIFIER TROUBLESHOOTING

Indicators/Observations	Possible Cause	Check or Monitor	Possible Solutions
1. Sludge floating to surface.	1a. "Bulking sludge" caused by filamentous organisms in the mixed liquor.	1a. Check Sludge Volume Index (SVI). If less than 100 then filaments are not the cause. SVI over 100 indicates filaments.	1a1. Increase D.O. in aeration tanks to greater than 1.0 mg/L. 1a2. Increase pH to 7.0. 1a3. Adjust nutrient balance so that BOD to nutrients is no more than 100 mg/L BOD to 5 mg/L total nitrogen, to 1 mg/L phosphorous, and 0.5 mg/L iron. 1a4. Add 10 to 50 mg/L chlorine or 50 to 200 mg/L hydrogen peroxide until SVI drops below 150.

Indicators/Observations	Possible Cause	Check or Monitor	Possible Solutions
			1a5. Increase sludge retention time (SRT).
			1a6. Increase sludge return rate.
	1b. "Rising sludge" due to denitrification occurring in clarifiers. Nitrogen gas given off attaches to sludge particles.	1b. Check nitrate concentration in clarifier influent. Presence of NO_3 will cause denitrification.	1b1. Increase sludge return rate.
			1b2. Increase D.O. in aeration tank.
			1b3. Reduce SRT.
	1c. Sludge not removed often enough (septic sludge).	1c. Check sludge collection and removal rates.	1c. Increase speed of collection and frequency of removal.
	1d. Overaeration.	1d. Check D.O. in aeration tank.	1d. Reduce D.O. and/or turbulence in aeration tank.

Indicators/Observations	Possible Cause		Check or Monitor		Possible Solutions	
2. Floc in effluent.	2a.	Excessive turbulence.	2a.	Aeration basin agitation, D.O.	2a.	Reduce aeration turbulence.
	2b.	SRT too long.	2b.	Check MLSS.	2b.	Increase sludge wasting to reduce SRT.
	2c.	Anaerobic conditions in aeration tank.	2c.	Check D.O. in aeration tank.	2c.	Increase D.O. in aeration.
	2d.	Toxic influent.	2d.	Check for dead protozoa in mixed liquor with microscope.	2d.	Identify and stop toxic discharge. Reseed with freeze-dried bugs or sludge from another plant.
	2e.	Short-circuiting in clarifier.	2e.	Look for unlevel weirs allowing short-circuited flow. Dye test.	2e.	Level weirs.
	2f.	Recycle of anaerobic side stream.	2f.	Look for anaerobic conditions.	2f.	Correct anaerobic conditions or eliminate side stream recycle.

Indicators/Observations	Possible Cause		Check or Monitor		Possible Solutions	
	2g.	Blanket too high.	2g.	Check blanket level.	2g.	Reduce blanket level by increasing return sludge and/or increasing wasting.
	2h.	Hydraulic overload.	2h.	Check flow rates.	2h.	Reduce flow.
	2i.	Incorrect polymer dosage.	2i.	Check polymer type, concentration, and feed rate. Perform a jar test.	2i.	Make adjustments to polymer feed based on jar test results.
3. Short-circuiting of flow through clarifiers.	3a.	Excessive hydraulic loading.	3a.	Check flow velocity.	3a.	Place more clarifiers in service.
	3b.	Weirs not level.	3b.	Visual inspection of weirs.	3b.	Level weirs.
	3c.	Reduced retention time caused by accumulation of large solids and grit.	3c.	Visual inspection.	3c.	Remove excessive solids accumulation and check operation of grit chamber.
4. Weir fouling.	4.	Accumulation of biological solids and growths.	4.	Visual inspection.	4.	More frequent and thorough cleaning.

Indicators/Observations	Possible Cause	Check or Monitor	Possible Solutions
5. Plugging of sludge ports.	5a. Collection of heavy compacted material.	5a. Visual inspection.	5a. Loosen material manually or with water jets.
	5b. Low velocity in withdrawal lines.	5b. Check sludge withdrawal rates.	5b. Pump sludge more frequently.
6. Excess torque on sludge collector rake.	6. Excessive load on rake.	6. Check torque meter and motor temperature.	6. May require repair or replacement of broken parts, more frequent operation of rake and sludge removal, or draining tank to look for a large object restricting the rake.

Indicators/Observations	Possible Cause	Check or Monitor	Possible Solutions
7. Breakup of floc in clarifier.	7a. Toxic or acid wastes.	7a. Highly turbid effluent.	7a. Identify and control industrial discharge.
	7b. Anaerobic conditions in aeration tank.	7b. Check aeration tank D.O.	7b. Increase D.O. in aeration tank.
	7c. Aeration tank overload.	7c. Check aeration tank loading.	7c. Place aeration basins in service.
	7d. Nutrient deficiency.	7d. Check mixed liquor nitrogen and phosphorous levels.	7d. Supplement nutrient deficiency by chemical addition.
	7e. Agitation.	7e. Check aeration basin turbulence.	7e. Reduce aeration basin turbulence.

SECONDARY CLARIFIERS 89

Indicators/Observations	Possible Cause	Check or Monitor	Possible Solutions
8. Sludge blanket overflowing.	8a. Inadequate rate of sludge return.	8a. Check sludge return rate and depth of sludge blanket.	8a1. Place another sludge return pump on line if one is malfunctioning.
			8a2. Control sludge return rate to keep sludge blanket between 1 to 3 feet.
			8a3. Clean sludge line if plugged.
	8b. Unequal flow distribution.	8b. Check flow to each clarifier.	8b. Adjust valves or gates to distribute flow equally.
	8c. Peak flows are overloading the clarifiers.	8c. Check hydraulic overflow rates.	8c. Put another clarifier on line, expand plant, or provide flow equalization facilities.
9. "Billowing sludge."	9a. Hydraulic surges.	9a. Visually inspect.	9a. Eliminate or dampen surges.
	9b. Density currents.	9b. Check sludge blanket depth.	9b. Keep sludge blanket as low as possible.
	9c. Stirring by rake.	9c. Visual inspection.	9c. Reduce rake mechanism speed.

NOTES

12 NITRIFICATION

12.1. PROCESS OVERVIEW

- The operation of biological processes may be manipulated to convert ammonia nitrogen to nitrate nitrogen.

- The nitrogen that is in raw sewage may be biologically oxidized to nitrate after the carbonaceous oxygen demand is met. This can only occur if the proper aerobic conditions are maintained in the process.

- Nitrification may be accomplished in a one- or two-stage process.

- In single-stage nitrification, the carbon and nitrogen oxidation steps are combined in a single unit.

- In two-stage nitrification, the carbonaceous oxidation step is carried out first in a separate unit, followed by the nitrification step in another unit. This may result in much higher oxygen demand, greater alkalinity consumption, and poorer settling, however.

12.2. NITRIFICATION TROUBLESHOOTING

Indicators/Observations	Possible Cause	Check or Monitor	Possible Solutions
1. Decrease in pH of nitrification system, with loss of nitrification.	1a. Not enough lime added to balance alkalinity destruction during nitrification.	1a. Check alkalinity in effluent from nitrification step.	1a. Add more lime or sodium hydroxide if alkalinity is less than 20 mg/L.
	1b. Not enough lime to keep pH from dropping due to chemicals added to remove phosphorous.	1b. Check pH and alkalinity in the influent to the nitrification step.	1b. Add more lime if pH is too low or alkalinity is less than 50 mg/L.
	1c. Acidic wastes in wastewater.	1c. Check pH and alkalinity in raw wastewater.	1c. Identify and stop acidic discharge.
	1d. Toxicants.	1d. Check for toxic influent.	1d. Identify and eliminate toxic source.

NITRIFICATION 93

Indicators/Observations	Possible Cause	Check or Monitor	Possible Solutions
2. Unable to completely nitrify. Effluent ammonia level too high.	2a. D.O. is too low.	2a. D.O. for nitrification must be 2.0 mg/L or more.	2a. Increase aeration or decrease loading in nitrification tank.
	2b. Cold temperatures.	2b. Check nitrification rate and MLVSS concentration.	2b. Decrease loading on nitrification tank or increase biological population in nitrification tank.
	2c. Increase in total daily influent nitrogen loads have occurred.	2c. Current influent nitrogen concentrations.	2c. Place additional nitrification units on line or modify pretreatment to remove more nitrogen.
	2d. SRT in nitrification unit too low.	2d. Check SRT.	2d1. Decrease loading on nitrification unit.
			2d2. Decrease wasting or loss of sludge from nitrification unit to increase SRT to 10 to 15 days (depending on temperature).

Indicators/Observations	Possible Cause	Check or Monitor	Possible Solutions
	2e. Peak hour ammonia concentration exceeds available oxygen supply.	2e. Check ammonia concentration trend.	2d3. Add settled raw sewage to nitrification unit to generate biological solids. 2e. Install flow equalization system to minimize peak concentrations or increase oxygen.
3. In two-stage system, SVI of nitrification sludge is too high (greater than 250).	3. Nitrification is occurring in the first stage.	3. Check for nitrates in first stage effluent.	3. Transfer sludge from first stage to second and maintain lower SRT in first stage.

NOTES

13 DENITRIFICATION

13.1. PROCESS OVERVIEW

- If bacteria come in contact with a nitrified element in the absence of oxygen, the nitrates are reduced to nitrogen gas (denitrification) and escape the wastewater.

- The denitrification process can be done in either an anoxic activated sludge system (suspended growth) or in a column system (fixed film).

- After effective biological treatment there is little oxygen demanding material left in the wastewater when it reaches the denitrification process. The denitrification reaction will only occur if there is an oxygen demand source when no dissolved oxygen is present in the wastewater.

- An oxygen demand source is usually added to reduce the nitrates quickly. The most common demand source added is soluble BOD or methanol. Approximately 3 mg/L of methanol is added for every 1 mg/L of nitrate-nitrogen.

- Suspended growth denitrification reactors are mixed mechanically, but only enough to keep the biomass from settling without adding unwanted oxygen.

- Submerged filters of different types of media may also be used to provide denitrification. A fine media, downflow filter is sometimes used to provide both denitrification and effluent filtration.

- A fluidized sand bed where wastewater flows upward through a media of sand or activated carbon at a rate to fluidize the bed may also be used. The denitrifying bacteria grow on the media.

13.2. DENITRIFICATION TROUBLESHOOTING

Indicators/Observations	Possible Cause	Check or Monitor	Possible Solutions
1. Effluent BOD increases.	1. Too much methanol or other oxygen demand material has been added.	1. Check methanol to nitrate-nitrogen ratio. (Should be 3 to 1).	1. Reduce methanol addition or improve methanol feed control.
2. Effluent nitrates increase.	2a. Not enough methanol being added.	2a. Check methanol feed system.	2a. Correct methanol feed.
	2b. pH is out of the desired 7.0 to 7.5 range due to low pH in the nitrification stage.	2b. Check alkalinity in nitrification stage effluent. Should be less than 30 mg/L.	2b. Add alkalinity (lime or sodium hydroxide) to the nitrification stage.
	2c. Loss of solids from denitrification stage due to failure of sludge return.	2c. Check denitrification stage solids accumulation and clarifier effluent.	2c. Increase sludge return. Decrease sludge wasting. Transfer sludge from carbonaceous unit to denitrifier.
	2d. Excessive mixing is bringing in D.O.	2d. Check denitrification stage D.O. (Should be less than 0.5 mg/L).	2d. Turn mixers down or off to reduce turbulence.

DENITRIFICATION 99

Indicators/Observations	Possible Cause	Check or Monitor	Possible Solutions
3. High headloss across filter bed type denitrification units.	3a. Excessive solids accumulation on or in filter bed.	3a. Check length of filter run. Most filters need backwashing after 12 hours of service.	3a. Initiate a full, complete backwash. Should also backwash before starting if the filter has been out of service for 24 hours or more.
	3b. Nitrogen gas accumulating and binding the filter.	3b. Check length of filter run. Probably nitrogen gas is the cause if the filter has less than 12 hours of service.	3b. Backwash bed for a brief period (1 to 2 minutes) and return to service.

NOTES

14 EFFLUENT FILTRATION

14.1. PROCESS OVERVIEW

- The filtration process involves wastewater being passed through a filter media of sand, anthracite, or a layer of each to remove suspended and/or colloidal material.

- Filtration is used as a form of tertiary treatment following secondary biological treatment or after physical chemical treatment (coagulation/flocculation/sedimentation).

- Filtration effectively reduces turbidity and improves chlorine disinfection by reducing chlorine demand.

- Filtration may be provided by open, gravity filters or in enclosed, pressure filters.

- Wastewater is passed downward through the filter media. With time, the filter media becomes plugged with material removed from the wastewater and must be cleaned by reversing the flow (backwashing). This upward backwash flow rate must be sufficient to fluidize the media particles, scrub and resuspend the collected particles, and wash them from the bed. The backwash wastewaters are usually recycled back to the head of the plant.

14.2. EFFLUENT FILTRATION TROUBLESHOOTING

Indicators/Observations	Possible Cause	Check or Monitor	Possible Solutions
1. High headloss through filter bed.	1a. Filter needs backwashing.	1a. Headloss should not be allowed to exceed 8 feet.	1a. Remove filter from service and backwash it.
	1b. Filter has become "air bound." (Holds entrapped gases).	1b. Headloss should not exceed 8 feet. Check D.O. of filter influent. The cause of an increase should be identified and controlled.	1b. Remove filter from service and backwash slowly to avoid filter damage caused by released air.
2. High headloss through a filter immediately after backwash.	2a. Insufficient backwash velocity or time.	2a. Check initial headloss. Normal initial headloss is 1 to 2 feet.	2a. Backwash filter for longer periods of time and/or increase backwash flow velocity.
	2b. Inoperative surface wash or air scour system.	2b. Visually inspect surface wash operation.	2b. Repair surface wash system.
3. Quick filter surface buildup and subsequent headloss.	3a. Inadequate clarification before filtration.	3a. Inspect operation of clarification step.	3a. Improve pretreatment performance.
	3b. Excessive filter aid polymer dosage.	3b. Check polymer feed rate.	3b. Reduce or eliminate filter aid polymer.
	3c. Inadequate surface wash and/or backwash.	3c. Check how quickly headloss increases.	3c. Provide adequate surface wash and backwash.

EFFLUENT FILTRATION

Indicators/Observations	Possible Cause	Check or Monitor	Possible Solutions
4. Short filter runs.	4a. High headloss caused by surface buildup.	4a. Visual inspection of filter surface.	4a. Replace filter media with dual or mixed media.
	4b. Filter aid polymer dosage too high.	4b. Check filter aid dosage.	4b. Reduce filter aid polymer dosage.
	4c. Solids carryover to filter.	4c. Check SS concentration.	4c. Provide improved pretreatment (settling).
	4d. Surface wash system not working.	4d. Inspect operation of surface wash system.	4d. Repair surface wash system.
	4e. Surface wash cycle not operated long enough.	4e. Check length of surface wash cycle.	4e. Increase surface wash time.
5. Effluent turbidity is too high.	5a. Filter needs backwashing.	5a. Turbidity in excess of 1 NTU.	5a. Remove filter from service and backwash it.
	5b. Inadequate pretreatment or chemical dosage upstream.	5b. Improve biological treatment or perform a jar test to correct chemical coagulation treatment (if used).	5b. Improve pretreatment or correct chemical coagulant dosage.

WASTEWATER TREATMENT TROUBLESHOOTING AND PROBLEM SOLVING

Indicators/Observations	Possible Cause	Check or Monitor	Possible Solutions
6. Effluent turbidity breakthrough but headloss is low.	6a. Inadequate filter aid polymer dosage.	6a. Check for high filter turbidity.	6a. Increase polymer dosage.
	6b. Coagulant feed system malfunction.	6b. Check operation of chemical feeders.	6b. Correct operation of chemical feeders.
	6c. Change in coagulant demand.	6c. Run a jar test.	6c. Adjust coagulant dose based on jar test results.
7. Mud ball formation.	7. Inadequate backwash flow rate and surface wash.	7. Inspect drained filter for mud balls.	7. Increase backwash flow rate and improve surface wash. Drain filter and manually remove mud balls with hoses or rakes.
8. Displaced gravel layer (media rests on gravel).	8. Introduction of air into filter underdrain during backwash.	8. Inspect filter surface for evidence of gravel breakthrough.	8. If displacement is severe, filter media must be replaced.
9. Loss of media during backwash.	9a. Excessive backwash flow.	9a. Check backwash rate.	9a. Reduce backwash flow rate.
	9b. Excessive surface wash.	9b. Check surface wash operation.	9b. Reduce surface wash time. Remember to turn off surface wash 2 minutes before completion of back-

EFFLUENT FILTRATION   105

Indicators/Observations	Possible Cause	Check or Monitor	Possible Solutions
10. Air binding and subsequent early headloss.	10a. Filter influent containing dissolved oxygen at or near saturation levels being subjected to less than atmospheric pressure in filter.	10a. Check for rapid headloss.	10a. Backwash more frequently at reduced flow and time until D.O. condition passes.
	10b. Lowered operating water level or allowing filter to run dry.	10b. Inspect water level in filter.	10b. Maintain maximum water depth over filters.
11. Difficulty in cleaning filters during warm weather.	11. Decreased viscosity of backwash water due to higher water temperature.	11. Check water and air temperatures. (Begins to occur when air temperature exceeds 80°F.)	11. Increase backwash rate until desired bed expansion during backwash is achieved.

Indicators/Observations	Possible Cause		Check or Monitor		Possible Solutions	
12. Percentage of backwash water used exceeds 5 percent of total flow.	12a.	Solids carryover to filter.	12a.	Check SS concentration.	12a.	Provided better pretreatment (improved settling).
	12b.	Filter aid polymer dosage too high.	12b.	Check filter aid polymer dosage.	12b.	Reduce filter aid polymer dosage.
	12c.	Surface wash system not working.	12c.	Inspect operation of surface wash system.	12c.	Repair surface wash system.
	12d.	Surface wash cycle not long enough.	12d.	Check length of surface wash cycle.	12d.	Increase surface wash cycle time.
	12e.	Filters are backwashed longer than necessary.	12e.	Check length of backwash.	12e.	Reduce the length of the backwash cycle.
13. Media surface sealing.	13.	Filter aid polymer over dosage.	13.	Check filter aid polymer dosage. Check filter media for evidence of polymer buildup.	13.	Reduce or eliminate filter aid polymer feed. If buildup is evident on media, media will need to be cleaned.

EFFLUENT FILTRATION

Indicators/Observations	Possible Cause	Check or Monitor	Possible Solutions
14. Media boils or is lost during backwash.	14. Backwash water is channeling up through areas in the filter rather than being distributed evenly. Backwash rate too high. Broken backwash water distribution system.	14. Check surface of filter for sand boils during backwash. Measure from top of sand to top of backwash troughs to determine if media is being lost.	14. Plugged or broken openings in underdrain system may allow channeling. This requires repair. Reduce backwash rate. Replace lost media.
15. Filter effluent turbidity increases. Chlorine demand increases.	15. Improper chemical pretreatment.	15. Assess overall process performance. Perform jar test if indicated.	15. Adjust coagulant dosage, optimize pretreatment. Decrease filtration rate, or start filter aid polymer feed. Increase chlorine dosage.

NOTES

15 CARBON ADSORPTION

15.1. PROCESS OVERVIEW

- The main purpose of carbon adsorption is the removal of soluble organic material that is difficult to remove by biological or physical/chemical treatment.

- The wastewater passes through a container filled either with carbon granules or a carbon slurry. Organics are removed by adsorption onto the carbon when there is sufficient contact time.

- A carbon system usually has several columns or basins used as contactors. Most contact chambers are either open concrete gravity-type systems, or steel pressure containers applicable to either upflow or downflow operation.

- With use, the carbon loses its adsorptive capacity. The carbon must then be regenerated or replaced with fresh carbon.

- Carbon used for adsorption at wastewater treatment plants may be in a granular form or in a powdered form.

- Granular carbon adsorption is accomplished by passing the wastewater through beds of the carbon housed in columns. These carbon columns usually provide 20 to 40 minutes of contact time.

- As headloss develops in carbon contactors, they are backwashed with clean effluent in much the same way the effluent filters are backwashed.

- Powdered carbon is too fine to use in columns and is usually added to the wastewater and then later removed by coagulation and settling.

15.2. CARBON ADSORPTION TROUBLESHOOTING

Indicators/Observations	Possible Cause	Check or Monitor	Possible Solutions
1. Excessive headloss through contactor.	1a. Highly turbid water applied to contactor.	1a. Check turbidity and SS concentration.	1a. Backwash to restore headloss. Improve pretreatment.
	1b. Growth and accumulation of biological solids in the contactor.	1b. Inspect carbon for growth. Check for hydrogen sulfide odor in effluent.	1b. Backwash or flush more frequently and improve upstream removal of soluble BOD.
	1c. Accumulation of carbon fines (small particles).	1c. Evidence of small particles.	1c. Remove carbon and wash out fines. Carbon may need to be replaced with a harder carbon.
	1d. Inlet or outlet screens plugged.	1d. Inspect screens.	1d. Backflush screens.

CARBON ADSORPTION 111

Indicators/Observations	Possible Cause	Check or Monitor	Possible Solutions
2. Hydrogen sulfide in contactor.	2a. Low D.O. and/or nitrate in contactor influent.	2a. Check influent D.O. and nitrate level.	2a. Add oxygen to influent.
	2b. High BOD concentration in influent.	2b. Check BOD in influent.	2b. Maintain aerobic conditions in carbon contactor/column. Improve upstream soluble BOD removal.
	2c. Excessive contact times.	2c. Check contact time (detention time). Should not exceed 60 minutes.	2c. Reduce contact time by removing one or more contactor or column from service.
3. Decrease in COD removal per pound of carbon.	3. Carbon exhausted and losing efficiency.	3. Check COD removal efficiency (COD removed per pound of carbon).	3. Replace or regenerate carbon.

NOTES

16 CHLORINATION

16.1. PROCESS OVERVIEW

- The most common use of chlorine in wastewater treatment is for disinfection. Other uses include odor control and activated sludge bulking control.

- Chlorine may also be used for nitrogen removal, through a process called breakpoint chlorination. For nitrogen removal, enough chlorine is added to the wastewater to convert all the ammonium nitrogen to nitrogen gas. To do this, approximately 10 mg/L of chlorine must be added for every 1 mg/L of ammonium nitrogen on the wastewater.

- For disinfection, chlorine is fed manually or automatically into a chlorine contact tank where it contacts flowing wastewater for at least 30 minutes to destroy disease causing microorganisms (pathogens) found in treated wastewater.

- Chlorine may be applied either as a gas or as a solid or liquid hypochlorite form.

16.2. CHEMICAL DECHLORINATION PROCESS OVERVIEW

- Dechlorination is the removal of all traces of residual chlorine remaining after the disinfection process and prior to the discharge of the final effluent to the receiving waters.

- This is commonly accomplished by the use of sulfur compounds such as sulfur dioxide, sodium sulfite, or sodium metabisulfite.

- Sulfur dioxide is the most popular method for dechlorination because it uses existing chlorination equipment.

Note: Because of their similarities, use the following chlorination troubleshooting section to diagnose and respond to sulfonator problems.

16.3. CHLORINATION TROUBLESHOOTING

Indicators/Observations	Possible Cause	Check or Monitor	Possible Solutions
1. Loss of chlorine gas pressure at chlorinator.	1a. Out of chlorine.	1a. Check chlorine cylinder.	1a. Switch to another container.
	1b. Cylinder valve not open.	1b. Check cylinder valve position.	1b. Open cylinder valve.
	1c. Plugged filter or pressure reducing valve.	1c. Check filter and reducing valve.	1c. Clean filter and/or repair reducing valve.
2. Low chlorine gas pressure at chlorinator.	2a. Insufficient number of cylinders connected.	2a. Reduce feed rate. Pressure will rise when you reduce the feed rate if there are not enough cylinders connected.	2a. Connect enough cylinders so the feed rate does not exceed the allowable withdrawal rate.
	2b. Flow restriction between cylinders and chlorinators.	2b. Reduce feed rate and look for ice formation.	2b. Disassemble chlorine header system at point where icing was noted, locate restriction and clean out or repair.

CHLORINATION

Indicators/Observations	Possible Cause	Check or Monitor	Possible Solutions
3. Chlorinator will not feed chlorine.	3a. Dirty pressure reducing valve in chlorinator.	3a. Inspect reducing valve.	3a. Disassemble chlorinator and clean valve stem and seat.
	3b. Loss of injector vacuum.	3b. Inspect injector.	3b. Repair injector. Clean piping.
	3c. No chlorine.	3c. Check chlorine supply.	3c. Renew chlorine supply.
	3d. Closed valve.	3d. Check valve positions.	3d. Open valve.
4. Loss of vacuum.	4a. Plugged diffuser	4a. Check diffuser.	4a. Clean and repair diffuser.
	4b. Injector failure.	4b. Check injector.	4b. Repair injector.
	4c. Loss of water pressure.	4c. Check water supply.	4c. Restore water pressure.
	4d. Vacuum leak in chlorinator.	4d. Check rotometer tube, leaky gasket, or ruptured diaphragm for leak.	4d. Locate and stop leak.

Indicators/Observations	Possible Cause	Check or Monitor	Possible Solutions
5. Abnormally high vacuum.	5. Loss of chlorine supply.	5. Check chlorine supply.	5. Restore chlorine supply.
6. Chlorine gas leak.	6. Numerous possible locations and causes.	6. Check for location of leak with an ammonia bottle (will form a white cloud at leak). Wear SCBA.	6. Locate, stop, and repair cause of gas leak.
7. Cold or freezing areas on supply system.	7a. Restriction in feed line.	7a. Look for cold spot to identify restricted area.	7a. Repair restriction.
	7b. Withdrawal rate too high.	7b. Compare withdrawal capacity with withdrawal rate.	7b. Reduce withdrawal rate or place more chlorine containers on line.

CHLORINATION 117

Indicators/Observations	Possible Cause	Check or Monitor	Possible Solutions
8. Liquid through manometer, or freezing of manometer.	8a. Defective evaporator.	8a. Check evaporator operation.	8a. Repair evaporator.
	8b. Feed rate too high.	8b. Check feed rate.	8b. Lower feed rate or put another chlorinator on line.
	8c. Restriction in manometer orifice.	8c. Check piping.	8c. Clean piping.
9. Misting.	9a. Defective evaporator.	9a. Check evaporator operation.	9a. Repair evaporator.
	9b. Undersized evaporator.	9b. Check feed rate.	9b. Put another chlorinator on line.
10. Low temperature alarm on evaporator.	10a. Dirty water bath.	10a. Check water bath.	10a. Clean water bath.
	10b. Loss of power.	10b. Check power.	10b. Restore power.
	10c. Heater element burned out.	10c. Check heater element.	10c. Replace heater element.

Indicators/Observations	Possible Cause	Check or Monitor	Possible Solutions
11. Inability to obtain maximum feed rate from chlorinator.	11a. Inadequate chlorine gas pressure.	11a. Check gas pressure.	11a. Increase pressure by adding or replacing cylinders.
	11b. Water injector clogged with deposits.	11b. Inspect injector.	11b. Clean injector.
	11c. Leak in vacuum relief valve.	11c. Disconnect vent line at chlorinator, place finger over vent connection to vacuum relief valve, observe if this increases vacuum and feed rate.	11c. Disassemble vacuum relief valve and replace all springs.
	11d. Vacuum leak in chlorinator system.	11d. Check system for leaks with ammonia solution bottle.	11d. Repair all vacuum leaks.

CHLORINATION 119

Indicators/Observations	Possible Cause	Check or Monitor	Possible Solutions
	11e. Faulty injector (no vacuum).	11e. Check injector system.	11e. Repair injector system.
	11f. Restriction in supply.	11f. Check for restrictions.	11f. Repair restrictions.
	11g. Leaks.	11g. Check for leaks.	11g. Repair leaks.
	11h. Wrong orifice.	11h. Check orifice.	11h. Install proper orifice.
12. Injector vacuum reading low.	12a. Dirty injector.	12a. Check injector water supply system.	12a. Clean injector.
	12b. Flow restricted.	12b. Check injector orifice.	12b. Adjust injector orifice.
	12c. Low pressure.	12c. Check throat.	12c. Close throat.
	12d. High pressure.	12d. Check throat.	12d. Open throat.
	12e. Back pressure.	12e. Check injector water supply system.	12e. Change injector and/or increase water supply to injector.

Indicators/Observations	Possible Cause	Check or Monitor	Possible Solutions
13. Leaking joints.	13. Missing gasket.	13. Check for missing gasket.	13. Repair joint. Replace gasket.
14. Chlorinator will not control low feed rates.	14a. Vacuum regulating valve.	14a. Check vacuum regulating valve and valve capsule.	14a. Repair diaphragm.
	14b. Chlorine pressure reducing valve (CPRV).	14b. Check CPRV.	14b. Clean CPRV cartridge, CPRV diaphragm, and CPRV gaskets.

16.4. CHLORINE CONTACT CHAMBER TROUBLESHOOTING

Indicators/Observations	Possible Cause	Check or Monitor	Possible Solutions
1. Increase in coliform level.	1a. Low chlorine residual.	1a. Check chlorine residual.	1a. Increase chlorine dosage.
	1b. Increased chlorine demand.	1b. Check BOD in influent to chamber.	1b. Improve upstream treatment and/or increase chlorine dosage.
	1c. Incomplete mixing.	1c. Check for short-circuiting.	1c. Baffle the chamber or install a mixer.
	1d. Solids accumulation in chamber.	1d. Check for solids in chamber.	1d. Take chamber out of service and clean.
	1e. Flow surges.	1e. Check flow patterns.	1e. Establish steady flow.
	1f. Plugged diffuser.	1f. Check diffuser.	1f. Clean diffuser.

Indicators/Observations	Possible Cause	Check or Monitor	Possible Solutions
2. Chlorine residual drops.	2a. Increased chlorine demand.	2a. Check BOD in influent to chamber.	2a. Improve upstream treatment and/or increase chlorine dosage.
	2b. Solids accumulation in chamber.	2b. Check for solids in chamber.	2b. Take chamber out of service and clean.
	2c. Change in flow rate.	2c. Check for flow rate increase.	2c. Adjust dosage for flow.
	2d. Feed equipment malfunction.	2d. Check chlorine feed equipment.	2d. Repair equipment.
	2e. Partial nitrification occurring.	2e. Check for nitrification.	2e. Correct process conditions to eliminate nitrification.
3. Chlorine residual increases.	3a. Decreased chlorine demand.	3a. Check BOD in influent to chamber.	3a. Lower chlorine dosage.
	3b. Change in flow rate.	3b. Check for flow rate decrease.	3b. Adjust dosage for flow.
	3c. Chlorine being overfed.	3c. Check chlorine feed rate.	3c. Decrease chlorine feed rate.

NOTES

17 GRAVITY THICKENING

17.1. PROCESS OVERVIEW

- Gravity thickening is a sedimentation process for thickening sludge. It is usually operated continuously with flow dependent on plant flow.

- The thickener is similar to a circular clarifier except that the bottom has more slope. Sludge enters at the center and solids settle to form a sludge blanket. The thickened sludge blanket is gently stirred by a rake mechanism to release gas bubbles, and to push the sludge to a center sump for removal. Supernatant passes over an effluent weir around the outer rim of the thickener.

- The supernatant flow is usually returned to either the primary treatment unit or to a secondary treatment process.

- The thickened sludge is usually pumped to either a blending, holding, or surge tank or directly to a dewatering process.

17.2. GRAVITY THICKENING TROUBLESHOOTING

Indicators/Observations	Possible Cause	Check or Monitor	Possible Solutions
1. Septic odor, rising sludge.	1a. Thickened sludge pumping is too low.	1a. Depth of sludge blanket (greater than 4 feet is too high).	1a. Increase underflow pumping of thickened sludge.
	1b. Thickener overflow rate is too low.	1b. Maintain an overflow rate of at least 600 gpd/sq ft.	1b. Increase influent flow (recirculate if necessary from secondary clarifier or add dilution water).
	1c. Septic wastewater.	1c. Check thickener D.O.	1c. Add chlorine or another oxidant to influent to provide 1.0 mg/L residual in the effluent.
	1d. Broken sludge collector rake.	1d. Check rake operation.	1d. Repair as necessary.
2. Thickened sludge too thin.	2a. Overflow rate is too high.	2a. Check overflow rate.	2a. Decrease influent pumping rate. Overflow rate should not exceed 800 gpd/sq ft.

GRAVITY THICKENING

Indicators/Observations	Possible Cause	Check or Monitor	Possible Solutions
	2b. Underflow withdrawal rate is too high.	2b. Check withdrawal pumping rate and sludge blanket level.	2b. Decrease withdrawal pumping rate. Keep blanket at least 2 to 3 feet deep.
	2c. Short-circuiting of flow through tank.	2c. Observe for a uneven discharge. Verify with a dye or tracer.	2c. Repair or relevel effluent weirs. Repair or reposition influent baffles.
3. Sludge collector torque overload.	3a. Heavy sludge accumulation.	3a. Check sludge blanket level in front of collector rake.	3a. Break up sludge in front of collector with rods or high pressure water hoses. Increase sludge withdrawal rate.
	3b. Heavy foreign object jamming the collector	3b. Probe in front of collector to locate object.	3b. Remove foreign object if possible with a hook or grappling device; or drain thickener and remove object.
	3c. Improper alignment.	3c. Check alignment.	3c. Realign collector mechanism.

Indicators/Observations	Possible Cause	Check or Monitor	Possible Solutions
4. Fine sludge particles in effluent.	4. Waste activated sludge (WAS).	4. Check percentage of WAS added to thickener.	4. Reduce WAS percentage or condition WAS before thickening.
5. Flow surges.	5. Influent pump cycles.	5. Check influent pump operation.	5. Modify pump cycling. Reduce flow and increase operation time.
6. Excessive growths on surfaces and weirs.	6. Inadequate cleaning.	6. Inspect surfaces.	6. More frequent cleaning or application of a weak chlorine solution.
7. Noisy or hot universal joint or bearings.	7. Excessive wear, improper alignment, or inadequate lubrication.	7. Check alignment and lubrication schedule.	7. Replace worn equipment, realign equipment, or lubricate as required.
8. Oil leak.	8. Seal failure.	8. Inspect oil seal.	8. Replace seal.
9. Pump overload.	9. Packing too tight, pump clogged, or sludge lines clogged.	9. Check packing gland for evidence of pump or line clog.	9. Adjust packing, clean pump and/or lines as required.

NOTES

18 DISSOLVED AIR FLOTATION (DAF) THICKENING

18.1. PROCESS OVERVIEW

- The flotation thickening process feeds air into the sludge to be thickened under pressure (40 to 80 psi), so that a large amount of air can be dissolved into the liquid.

- Air is dissolved into the wastewater, under pressure, in a retention tank by means of an air compressor. This air/water mixture is injected into and mixed with the sludge.

- The air saturated sludge then flows into an open tank where, at atmospheric pressure, much of the air comes out of solution as tiny bubbles that attach to sludge particles and float them to the surface.

- As additional solids rise, compaction occurs, resulting in a layer of thickened sludge or "blanket" at the surface of the tank.

- This thickened sludge is removed by skimmers that continually skim off the upper surface of the DAF tank sludge blanket.

- Part of the DAF effluent is pumped through the retention tank. Air is fed into the pump discharge or into the retention tank.

- The flow through this recycle system is controlled by a valve. Effluent recycle can range from 50 to 300 percent of the influent flow. This recycle flow and sludge feed are mixed in a chamber at the entrance to the DAF unit.

- Polymer, due to its ability to bridge individual sludge particles, is often used to improve DAF performance and solids capture. If a polymer flotation aid is used, it is usually fed to the mixing chamber.

- Clarified effluent flows under and over a weir, and bottom sludge collectors are used to remove any settled sludge or grit.

18.2. DISSOLVED AIR FLOTATION THICKENING TROUBLESHOOTING

Indicators/Observations	Possible Cause	Check or Monitor	Possible Solutions
1. Thickened sludge too thin.	1a. Skimmer speed too high.	1a. Observe skimmer operation.	1a. Adjust skimmer speed.
	1b. Unit is overloaded.	1b. Check rise rate.	1b. Stop feed sludge and allow unit to clear itself.
	1c. Polymer dosage too low.	1c. Check polymer dosage and polymer feed pump operation.	1c. Adjust dosage as needed.
	1d. Excessive air/solids ratio.	1d. Float blanket may appear frothy.	1d. Reduce air flow to pressure saturation system.
2. Low dissolved air (sludge float too thin).	2a. Reaeration pump off, malfunctioning, or clogged.	2a. Inspect pump condition and operation.	2a. Clean or repair as required.
	2b. Eductor clogged.	2b. Inspect eductor.	2b. Clean eductor.
	2c. Malfunction in air supply system.	2c. Check air compressor, air lines, and controls.	2c. Repair as required.

DISSOLVED AIR FLOTATION (DAF) THICKENING 133

Indicators/Observations	Possible Cause	Check or Monitor	Possible Solutions
3. Effluent solids too high.	3a. Unit is overloaded.	3a. Check rise rate.	3a. Stop feed sludge and allow unit to clear itself.
	3b. Polymer dosage is too low.	3b. Check polymer dosage and polymer feed pump operation.	3b. Adjust dosage as needed.
	3c. Skimmer off or too slow.	3c. Check skimmer operation.	3c. Adjust skimmer speed.
	3d. Low air/solids ratio.	3d. Poor float formation with settling solids.	3d. Increase air flow to pressure/saturation system.
4. Rise rate too slow.	4a. Unit is overloaded.	4a. Check rise rate.	4a. Stop feed sludge and allow unit to clear itself.
	4b. Low dissolved air.	4b. Reaeration pump, eductor, and air supply.	4b. Clean or repair components as required.
	4c. Polymer dosage too low.	4c. Check polymer dosage and polymer feed pump operation.	4c. Adjust dosage as needed.

Indicators/Observations	Possible Cause	Check or Monitor	Possible Solutions
5. Skimmer blade leaking at beach.	5a. Skimmer blade not adjusted properly.	5a. Inspect blade beach contact.	5a. Adjust as required.
	5b. Hold down tracks too tight.	5b. Inspect tracks.	5b. Adjust as required.
6. Skimmer blade binding on beach.	6. Skimmer blade not adjusted properly.	6. Inspect blade beach contact.	6. Adjust as required.
7. Water level in retention tank too high.	7a. Low air supply pressure.	7a. Check air compressor and air lines.	7a. Repair as required.
	7b. Level control system not working.	7b. Check operation of level control system.	7b. Repair as required.
	7c. Not enough air being injected.	7c. Check air compressor and air lines.	7c. Increase air or repair problem.
8. Water level in retention tank too low.	8a. Recirculation pump malfunction.	8a. Check pump operation.	8a. Repair or clean as required.
	8b. Level control system not working.	8b. Check operation of level control system.	8b. Repair as required.
9. Recirculation pump capacity is low.	9. Retention tank pressure too high.	9. Check back pressure.	9. Adjust back pressure valve.

NOTES

19 ANAEROBIC DIGESTION

19.1. PROCESS OVERVIEW

- In this process, the organic matter in sludge is broken down by microorganisms in the absence of oxygen.

- Anaerobic digestion is usually divided into two operational modes, either "high rate" or "low rate."

- In low rate, or one stage digestion, fresh sludge is fed into the digester two or three times daily. As digestion occurs, three distinct layers form in the tank. A scum layer is formed at the top of the digester, a supernatant level below the scum, and a sludge layer at the bottom. The sludge layer also has an actively decomposing top layer and a stabilized bottom layer.

- High rate systems may use one or two stages. A two-stage high rate system uses a first stage digester for sludge stabilization, and a second-stage digester for settling and thickening of the decomposed sludge. In a single-stage high rate system, the second stage is replaced by some other thickening process.

- Anaerobic digesters are heated to 85 to 95°F and are operated to provide 10 to 20 days detention time for the digesting sludge.

- Anaerobic digestion converts about 50 percent of the organic solids to liquid and gas. The gas can be used as fuel, the liquid phase is usually returned to the head of the plant, and the remaining sludge is disposed of.

- The process is most successful when fed a high percentage of primary sludge. Waste activated sludge does not settle or dewater well after digestion.

- Key factors in anaerobic digestion include the following concepts:

$$\text{Volatile solids loading,} \frac{\text{lbs VS/day}}{\text{cu ft}} = \frac{\text{feed sludge VS, lbs/day}}{\text{digester volume, cu ft}}$$

$$\text{Volatile solids reduced, percent} = \frac{\text{lbs VS in} - \text{lbs VS out}}{\text{lbs VS in}} \times 100$$

$$\text{Detention time, days} = \frac{\text{digester volume, gallons}}{\text{sludge feed, gals/day}}$$

$$\text{Gas production,} \frac{\text{cu ft gas}}{\text{lbs VS fed}} = \frac{\text{gas produced, cu ft per day}}{(\text{Vs fed, lbs/day})(\text{reduction}/100)}$$

19.2. ANAEROBIC DIGESTION TROUBLESHOOTING

Indicators/Observations	Possible Cause	Check or Monitor	Possible Solutions
1. A rise in the volatile acid to alkalinity ratio.	1a. Hydraulic overload.	1a. Monitor volatile acids, alkalinity, and temperature until ratio stabilizes.	1a. If ratio increases to 0.3, add seed sludge from a healthy digester and decrease sludge withdrawal, extend mixing time, and assure that the digester temperature is steady.
	1b. Organic overload.	1b. Monitor sludge pumping volume and amount of volatile solids in feed sludge. Reduce or eliminate septic tank sludge or industrial wastes discharged to plant.	1b. If ratio increases to 0.3, add seed sludge from a healthy digester and decrease sludge withdrawal, extend mixing time, assure that the digester temperature is steady, and decrease loading rate.
	1c. Discharge of toxic materials to digester.	1c. Monitor volatile acids, pH, and gas production. Check for industrial waste dischargers or for presence of sulfides from inadequate sludge pumping.	1c. Recycle solids, dilute with liquid, decrease feed concentration, keep pH in digester above 7.0, and monitor industrial dischargers.

Indicators/Observations	Possible Cause	Check or Monitor	Possible Solutions
2. CO_2 content in gas increases.	2. Volatile acids to alkalinity ratio has increased to 0.5.	2a. Check waste gas flame color. 2b. Use a gas analyzer.	2a. Add alkalinity. 2b. Add alkalinity.
3. pH drops. CO_2 increases to a level where gas burner goes out.	3. Volatile acids to alkalinity ratio has increased to 0.8.	3a. Monitor gas burner and analyze gas. 3b. Check for rotten egg hydrogen sulfide odor.	3a. Add alkalinity. 3b. Decrease loading rate to less than 0.01 lb. volatile solids per cu ft per day until ratio drops below 0.5.

Indicators/Observations	Possible Cause	Check or Monitor	Possible Solutions
4. Supernatant quality is too strong to return to head of plant without process upset.	4a. Too much mixing and not enough settling time.	4a. Withdraw a sample and observe settling characteristics.	4a. Allow a longer period for settling before drawing off supernatant.
	4b. Supernatant draw-off piping not at the same level as the supernatant layer.	4b. Locate the supernatant layer by withdrawing samples at different levels.	4b. Adjust tank operating level or supernatant draw-off pipe.
	4c. Raw sludge feed point too close to supernatant draw-off line.	4c. Determine volatile solids content. Should be close to that of well mixed sludge and much lower than raw sludge.	4c. Revise piping when digester is next dewatered.
	4d. Not withdrawing enough digested sludge.	4d. Compare feed and withdrawal rates. Check volatile solids to see if sludge is fully digested.	4d. Increase digested sludge withdrawal rates. Note: do not withdraw more than 5 percent of digester volume per day.

Indicators/Observations	Possible Cause	Check or Monitor	Possible Solutions
5. Supernatant from either primary or secondary digester has a sour smell.	5a. The pH of digester is too low.	5a. Refer to 3 above.	5a. Refer to 3 above.
	5b. Overloaded digester ("rotten egg odor").	5b. Refer to 3 above.	5b. Refer to 3 above.
	5c. Toxic load.	5c. Monitor volatile acids, pH, and gas production. Check industrial discharges and for presence of sulfides.	5c. Recycle solids, dilute with liquid, decrease feed concentration, and keep pH in digester above 7.0.
6. Foam observed in supernatant from single stage or primary tank.	6a. Scum blanket is breaking up.	6a. Check condition of scum blanket.	6a. Normal condition, but stop withdrawing supernatant for a while if possible.
	6b. Excessive gas recirculation.	6b. Do not exceed 20 cfm/1,000 cu ft.	6b. Throttle compressor output.
	6c. Organic overload.	6c. Volatile solids loading ratio.	6c. Reduce feeding rate.

ANAEROBIC DIGESTION   143

Indicators/Observations	Possible Cause	Check or Monitor	Possible Solutions
7. Bottom sludge too watery or too thin.	7a. Short-circuiting.	7a. Draw-off line to supernatant zone is open.	7a. Change to draw-off line at the bottom of digester.
	7b. Too much mixing.	7b. Take sample and check how it settles.	7b. Shut off mixing for 24 to 48 hours before drawing sludge.
	7c. Sludge coning, allowing lighter solids to be pulled into pump suction.	7c. Total solids test or visual observation.	7c. "Bump" the pump two to three times by starting and stopping, pump digester contents back through the withdrawal line, or attach a water hose to the pump suction line and force water through it. (Water source must be non-potable). Run for only 2 to 3 minutes to avoid diluting the digester with water.

Indicators/Observations	Possible Cause	Check or Monitor	Possible Solutions
8. Sludge temperature is falling and can not be maintained.	8a. Sludge is plugging heat exchanger.	8a. Check inlet and outlet pressure of exchanger.	8a. Open heat exchanger and clean.
	8b. Sludge recirculation line is partially or completely plugged.	8b. Check pump inlet and outlet pressure.	8b. Backflush the line with heated digester sludge, use mechanical cleaner, apply water pressure, or add approximately 3 lbs/100 gallons of water of trisodium phosphate (TSP) or other commercial degreaser.
	8c. Not enough mixing.	8c. Check temperature profile in digester.	8c. Increase mixing.
	8d. Hydraulic overload.	8d. Check incoming sludge concentration.	8d. If ratio increases to 0.3, add seed sludge and decrease sludge withdrawal, extend mixing time, and assure that the digester temperature is steady.
	8e. Low water feed rate in internal coils used for heat exchange.	8e. Air lock in line or a valve partially closed.	8e. Bleed air relief valve.

ANAEROBIC DIGESTION   145

Indicators/Observations	Possible Cause	Check or Monitor	Possible Solutions
	8f. Boiler burner not firing on digester gas.	8f. Low gas pressure or unburnable gas due to process upset.	8f. Locate and repair leak. Also see 3 above.
	8g. Heating coils inside digester are coated.	8g. See if the temperature of inlet and outlet water is about the same.	8g. Remove coating (may require draining tank) and control water temperature to 130°F maximum.
9. Sludge temperature is rising.	9. Temperature controller is not working properly.	9. Check water temperature and controller setting.	9. If over 120°F, reduce temperature. Repair or replace controller.
10. Recirculation pump not running; power circuits O.K.	10. Temperature override in circuit to prevent pumping too hot water through tubes.	10. Visual check, no pressure on sludge line.	10. Allow system to cool off. Check temperature control circuits.
11. Gas mixer feed lines plugging.	11a. Lack of flow through gas lines.	11a. Identify low temperature of gas feed pipes or low pressure.	11a. Flush out with water.
	11b. Debris in gas lines.	11b. Check for low temperature of gas feed pipes or low pressure.	11b. Clean feed lines and/or valves, and provide thorough service when tank is drained for inspection.

Indicators/Observations	Possible Cause	Check or Monitor	Possible Solutions
12. Gear reducer wear on mechanical mixers.	12a. Lack of proper lubrication.	12a. Excessive motor amperage, excessive noise and vibration, evidence of shaft wear.	12a. Verify correct type and amount of lubrication from manufacturer's literature.
	12b. Misalignment of equipment.	12b. Refer to 15 below.	12b. Correct imbalances caused by accumulation of material on the internal moving parts.
13. Shaft seal leaking on mechanical mixer.	13. Packing dried out or worn.	13. Evidence of gas leakage (odor of gas).	13. Follow manufacturer's instructions for repacking. Note: replace packing any time the tank is empty if it is not possible when unit is operating.
14. Wear on internal parts of mechanical mixer.	14. Grit or misalignment.	14. Visual observation when tank is empty, compare with manufacturer's drawings of original size. Note: motor amperage will go down as moving parts are worn away and get smaller.	14. Replace or rebuild parts as necessary.

ANAEROBIC DIGESTION 147

Indicators/Observations	Possible Cause	Check or Monitor	Possible Solutions
15. Imbalance of internal parts due to accumulation of debris on the moving parts of mechanical mixers.	15. Poor comminution and/or screening.	15. Vibration, motor heating, excessive amperage or noise.	15. Reverse direction of mixer, stop and start, open inspection hole for a visual inspection, or draw tank down and clean moving parts.
16. The desired rolling movement of scum blanket is slight or absent.	16a. Mixer is off.	16a. Mixer switch or timer.	16a. May be normal if mixers are set on a timer. If not and mixers should be operating, check for malfunction.
	16b. Inadequate mixing.	16b. Mixer operation.	16b. Increase mixing.
	16c. Scum blanket is too thick.	16c. Measure blanket thickness.	16c. Refer to 18 and 19 below.
17. Scum blanket is too high.	17. Supernatant overflow line is plugged.	17. Check gas pressure; it may be above normal or relief valve may be venting to atmosphere.	17. Lower contents through bottom draw-off, then rod supernatant line to clear plugging.

148 WASTEWATER TREATMENT TROUBLESHOOTING AND PROBLEM SOLVING

Indicators/Observations	Possible Cause	Check or Monitor	Possible Solutions
18. Scum blanket is too thick.	18. Lack of mixing, high grease content.	18. Probe blanket for thickness through thief hole or in gap beside floating cover.	18. Break up blanket by using mixers, use sludge recirculation pumps and discharge above the blanket, use chemicals to soften blanket, or break up blanket physically with pole.
19. Draft tube mixers not moving surface adequately.	19. Scum blanket too high and allowing thin sludge to travel under it.	19. Rolling movement on sludge surface.	19. Lower sludge level to 3 to 4 inches above top of tube, allowing thick material to be pulled into tube - continue for 24 to 48 hours, or reverse direction (if possible).
20. Gas is leaking through pressure relief valve (PRV) on roof.	20. Valve not seating properly or is stuck open.	20. Check to see if digester gas pressure is normal.	20. Remove PRV cover and move weight holder until it seats properly. Install new ring if needed.

ANAEROBIC DIGESTION

Indicators/Observations	Possible Cause	Check or Monitor	Possible Solutions
21. Digester gas pressure is above normal.	21a. Obstruction in water in main burner gas line.	21a. If all use points are operating and normal, then check for a waste gas line restriction or a plugged or stuck safety device.	21a. Purge with air, drain condensate traps, check for low spots. Care must be taken not to force air into digester.
	21b. Digester PRV is stuck shut.	21b. Gas is not escaping as it should.	21b. Remove PRV cover and manually open valve, clean valve seat.
	21c. Waste gas burner line pressure control valve is closed.	21c. Gas meters show excess gas is being produced, but not going to waste gas burner.	21c. Re-level floating cover if gas escapes around dome due to tilting.
22. Digester gas pressure below normal.	22a. Fast withdrawal causing a vacuum inside digester.	22a. Check vacuum breaker to be sure it is operating properly.	22a. Stop supernatant discharge and close off all gas outlets from digester until pressure returns to normal.
	22b. Too much lime has been added.	22b. Sudden increase in CO_2 in digester gas.	22b. Stop addition of lime and increase mixing.

150 WASTEWATER TREATMENT TROUBLESHOOTING AND PROBLEM SOLVING

Indicators/Observations	Possible Cause	Check or Monitor	Possible Solutions
23. Pressure regulating valve not opening as pressure increases.	23a. Inflexible diaphragm.	23a. Isolate valve and open cover. Check for leaks.	23a. If no leaks are found diaphragm may be lubricated and softened using neats-foot oil.
	23b. Ruptured diaphragm.	23b. Visual inspection.	23b. Replace diaphragm.
24. Yellow gas flame from waste gas burner.	24. Poor quality gas with a high CO_2 content.	24. Check CO_2 to see if content is higher than normal.	24. Check concentration of sludge feed (may be too dilute). If dilute, increase sludge concentration.
25. Gas meter failure (propeller or lobe type).	25a. Trash in line.	25a. Condition of gas line.	25a. Flush with water, isolating digester and working from digester out toward points of usage.
	25b. Mechanical failure.	25b. Fouled or worn parts.	25b. Clean with kerosene and/or replace worn parts.

ANAEROBIC DIGESTION 151

Indicators/Observations	Possible Cause	Check or Monitor	Possible Solutions
26. Gas meter failure (bellows type).	26a. Inflexible diaphragm.	26a. Isolate valve, open cover, and check for leaks.	26a. If no leaks are found, diaphragm may be lubricated and softened using neatsfoot oil.
	26b. Ruptured diaphragm.	26b. Visual inspection.	26b. Replace diaphragm. (Metal guides may also need to be replaced if corroded.)
27. Gas pressure higher than normal during freezing weather.	27a. Supernatant line plugged.	27a. Supernatant overflow lines.	27a. Check every 2 hours during freezing conditions, inject steam, protect line from weather by covering and insulating overflow box.
	27b. Pressure relief valve stuck or closed.	27b. Weights on pressure relief valves.	27b. If freezing is a problem, apply light grease and rock salt layer.

Indicators/Observations	Possible Cause	Check or Monitor	Possible Solutions
28. Gas pressure lower than normal.	28a. Pressure relief valve or other pressure control devices stuck open.	28a. Pressure relief valve and devices.	28a. Manually operate vacuum relief and remove corrosion.
	28b. Gas line or hose leaking.	28b. Gas line and/or hose.	28b. Repair or replace as needed.
29. Leaks around metal covers.	29. Anchor bolts pulled loose and/or sealing material moved or cracking.	29. Check for concrete broken around anchors, tie-downs bent, sealing materials displaced.	29. Repair concrete with fast sealing concrete repair material. New tie-downs may have to be welded onto old ones and re-drilled. (Tanks must be drained and well ventilated before any welding). New sealant material should be applied to any leak.

ANAEROBIC DIGESTION 153

Indicators/Observations	Possible Cause	Check or Monitor	Possible Solutions
30. Gas suspected leaking through concrete cover.	30. Freezing and thawing causing widening of construction cracks.	30. Check for gas leaks by applying a soapy solution and looking for bubbles.	30. If serous leaks are identified, drain tank, clean cracks and repair with concrete sealers. (Tanks must be drained and well ventilated for this procedure.)
31. Floating cover tilting, little or no scum around the edges.	31a. Weight distributed unevenly.	31a. Location of weights.	31a. If moveable ballast or weights are provided, relocate them until the cover is level. If no weights are provided, use sandbags to level the cover. (Note: pressure relief valves may need to be reset if significant weight is added.)
	31b. Water from condensation or rain water collecting on top of metal cover in one location.	31b. Check around the edges of the metal cover. (Some covers with insulating wooden roofs have inspection holes for this purpose.)	31b. Remove the water. Repair roof if leaks in the roof are contributing to the water problem.

154 WASTEWATER TREATMENT TROUBLESHOOTING AND PROBLEM SOLVING

Indicators/Observations	Possible Cause	Check or Monitor	Possible Solutions
32. Floating cover tilting, heavy thick scum accumulating around edges.	32a. Excess scum in one area, causing excess drag.	32a. Probe with a stick or some other method to determine the condition of the scum.	32a. Use chemicals or degreasing agents to soften the scum, then hose down with water.
	32b. Guides or rollers out of adjustment.	32b. Distance between guides or rollers and the wall.	32b. Soften up the scum and readjust rollers for guides so that skirt doesn't rub on the walls.
	32c. Rollers or guides broken.	32c. Determine the normal position if the suspected broken part is covered by sludge. Verify correct location using manufacturer's information and/or drawings.	32c. Drain tank if necessary. Take care as cover lowers to corbels so it does not bind or come down unevenly. It may be necessary to use a crane or jacks in order to prevent structural damage.

Indicators/Observations	Possible Cause	Check or Monitor	Possible Solutions
33. Cover binding even though rollers and guides are free.	33. Internal guide or guy wires are binding or damaged (some covers are built like umbrellas with guides attached to the center column).	33. Lower down to corbels. Open hatch and inspect from the top using self-contained breathing apparatus and an explosion-proof light. If cover will not go all the way down, it may be necessary to secure in one position with a crane or by other means to prevent skirt damage to sidewalls.	33. Drain and repair, holding the cover in a fixed position if necessary.

NOTES

20 AEROBIC DIGESTION

20.1. PROCESS OVERVIEW

- Aerobic digestion is the separate aeration of primary sludge, waste biological sludge, or a blend of these two in an open or closed tank.

- A separate sedimentation tank may be used following aerobic digestion, or a one tank, batch type system where the sludge is aerated, mixed, settled, and decanted in the same tank may be used.

- The aerobic digestion process is an extension of the extended aeration process. The volatile material in the wastes is digested to a reasonable maximum with up to a 45 percent destruction of volatile solids.

- The decomposition of solids and regrowth of organisms is maintained until the available energy in active cells and the storage of waste materials are sufficiently low and stable enough for disposal.

- Aerobic digestion reduces the volume of sludge solids and reduces odors or other nuisances that may be a hindrance to final disposal.

20.2. AEROBIC DIGESTION TROUBLESHOOTING

Indicators/Observations	Possible Cause	Check or Monitor	Possible Solutions
1. Excessive foaming.	1a. Organic overload.	1a. Organic load.	1a. Reduce feed rate, and increase solids in digester by decanting and recycling solids.
	1b. Excessive aeration.	1b. Dissolved oxygen.	1b. Reduce aeration rate.
2. Low dissolved oxygen.	2a. Diffusers clogging.	2a. Dewater digester, withdraw sludge and inspect diffusers.	2a. Clean or replace diffusers.
	2b. Liquid level not correct for mechanical aeration.	2b. Check equipment specifications.	2b. Establish correct liquid level.
	2c. Blower malfunction.	2c. Air delivery rate, pipeline pressure valving.	2c. Repair pipe leaks, set valves in proper position, repair blower.
	2d. Organic overload.	2d. Organic load.	2d. Reduce feed rate, and increase solids in digester by decanting and recycling solids.

AEROBIC DIGESTION 159

Indicators/Observations	Possible Cause	Check or Monitor	Possible Solutions
3. Sludge has unpleasant odor.	3a. Inadequate SRT.	3a. Check SRT.	3a. Reduce feed rate, and increase solids in digester by decanting and recycling solids.
	3b. Inadequate aeration.	3b. D.O. should be greater than 1 mg/L.	3b. Increase aeration or reduce feed rate.
4. Ice formation damages mechanical aerators.	4. Extended freezing weather.	4. Check digester surface for ice information.	4. Break up and remove ice before it causes damage.
5. pH in digester has dropped to undesirable level (below 6.0 to 6.5).	5a. Nitrification is occurring and wastewater alkalinity is low.	5a. pH of supernatant.	5a. Add sodium bicarbonate to feed sludge or lime or sodium hydroxide to digester.
	5b. CO_2 is accumulating in covered digester in air space and is dissolving into sludge.	5b. pH of supernatant.	5b. Vent and scrub the digester gas.

NOTES

21 CENTRIFUGATION

21.1. PROCESS OVERVIEW

- Solid/liquid separation occurs in a centrifuge as a result of rotating the liquid at high speeds to cause separation by gravity.

- There are many types, but the most common centrifuge for dewatering sewage sludge is the solid bowl centrifuge. The solid bowl type has a rotating unit with a bowl and a conveyor. The unit has a conical section at one end that acts as a drainage device. The conveyor screw pushes the sludge solids to outlet ports and the cake to a discharge hopper.

- The sludge slurry enters the rotating bowl through a feed pipe leading into the hollow shaft of the rotating screw conveyor. The sludge is distributed through ports into a pool inside the rotating bowl.

- As the liquid sludge flows through the hollow shaft toward the overflow devices, the finer solids settle to the wall of the rotating bowl. The screw conveyor pushes the solids to the conical section where the solids are forced out of the water, and the water drains back into the pool.

21.2. CENTRIFUGATION TROUBLESHOOTING

Indicators/Observations	Possible Cause	Check or Monitor	Possible Solutions
1. Poor centrate clarity.	1a. Feed rate too high.	1a. Flow records.	1a. Reduce flow.
	1b. Low pool depth (wrong dam setting).	1b. Pool depth and dam setting.	1b. Change dams to increase pool depth.
	1c. Worn conveyor flights.	1c. Vibration or solids buildup.	1c. Repair or replace conveyor.
	1d. Speed too high.	1d. Speed control.	1d. Reduce speed.
	1e. Feed solids too high.	1e. Spin test feed sludge.	1e. Dilute feed sludge (should be less than 40% by volume).
	1f. Incorrect chemical conditioning.	1f. Chemical feed rate.	1f. Adjust chemical dosage.
2. Cake too wet.	2a. Feed rate too high.	2a. Flow records.	2a. Reduce flow.
	2b. High pool (wrong dam setting).	2b. Pool depth and dam setting.	2b. Change dams to decrease pool depth.
	2c. Speed too low.	2c. Speed control.	2c. Increase speed.
	2d. Excessive chemical feed.	2d. Chemical feed rate.	2d. Decrease chemical dosage.

CENTRIFUGATION

Indicators/Observations	Possible Cause	Check or Monitor	Possible Solutions
3. Centrifuge high torque trips.	3a. Feed rate too high.	3a. Flow records.	3a. Reduce flow.
	3b. Feed solids too high.	3b. Spin test feed sludge.	3b. Dilute feed sludge (should be less than 40% by volume).
	3c. Foreign material in machine.	3c. Inspect interior.	3c. Remove conveyor and remove foreign material.
	3d. Misaligned gear unit.	3d. Vibration.	3d. Realign gear unit.
	3e. Gear unit has faulty gear, bearing, or spline.	3e. Inspect gear unit.	3e. Replace faulty components.

Indicators/Observations	Possible Cause	Check or Monitor	Possible Solutions
4. Excessive vibration.	4a. Improper lubrication.	4a. Check lubrication system.	4a. Provide correct lubrication.
	4b. Improper adjustment of vibration isolators.	4b. Vibration isolators.	4b. Adjust isolators.
	4c. Discharge funnels may be contacting centrifuge.	4c. Funnel positions.	4c. Adjust funnel position.
	4d. Portion of conveyor screw may be plugged with solids causing an imbalance.	4d. Machine interior.	4d. Flush centrifuge to clean out solids.
	4e. Improper gear box alignment.	4e. Gear box alignment.	4e. Realign gear box.
	4f. Damaged pillow block bearings.	4f. Inspect bearings.	4f. Replace bearings.
	4g. Bowl unbalanced.	4g. Check bowl balance.	4g. Balance rotating parts.
	4h. Loose parts.	4h. Check tightness.	4h. Tighten parts.
	4i. Uneven conveyor wear.	4i. Inspect conveyor.	4i. Resurface and rebalance conveyor.

Indicators/Observations	Possible Cause	Check or Monitor	Possible Solutions
5. Increased power usage.	5a. Bowl contacting accumulated solids in centrifuge cake.	5a. Look for signs of wear.	5a. Resurface worn areas. Remove accumulated solids.
	5b. Plugged effluent pipe.	5b. Check for free solids discharge.	5b. Clear effluent pipe.
	5c. Conveyor blade wear.	5c. Conveyor condition.	5c. Resurface blades.
6. Surging, spasmodic solids discharge.	6a. Pool depth too low (wrong dam settings).	6a. Check pool depth and dam setting.	6a. Increase pool depth.
	6b. Conveyor helix screw rough.	6b. Corrosion or improper hard surfacing.	6b. Rebuild/refinish conveyor screw.
	6c. Feed pipe too near bowl beach.	6c. Check location of feed pipe relative to bowl beach.	6c. Move feed pipe to effluent end.

Indicators/Observations	Possible Cause	Check or Monitor	Possible Solutions
7. Centrifuge won't start or shuts down after starting.	7a. Tripped breaker or blown fuse.	7a. Check breakers and fuse.	7a. Replace fuse or reset breaker and restart.
	7b. Overload relay tripped.	7b. Check overload switch.	7b. Flush machine, reset relay switch, and restart.
	7c. Motor overheated, thermal protection tripped.	7c. Thermal protectors.	7c. Flush machine, reset thermal protectors, and restart.
	7d. Torque control tripped.	7d. Check overtorque causes in No. 3.	7d. Correct problem and attempt a restart.
	7e. Vibration switch tripped.	7e. Check vibration causes in No. 4.	7e. Correct problem and attempt to restart.
	7f. Other interlocked equipment failure.	7f. Interlocked equipment.	7f. Reset.

NOTES

22 VACUUM FILTRATION

22.1. PROCESS OVERVIEW

- A vacuum filter is a device used to separate solid material from liquid.

- Sludge is chemically treated and placed in a vat or tank. The vacuum filter has a round drum which rotates partially submerged in the sludge.

- A vacuum is applied to the inside of the drum to draw the sludge onto the outside of the drum cover. This porous outside cover or filter medium allows the filtrate of liquid to pass through into the drum and the filter cake to stay on the medium.

- The filtrate water is returned to the plant for treatment and the filter cake solids pulled to the outside of the medium are removed by a fixed scraper blade and are discharged onto a conveyor belt.

22.2. VACUUM FILTRATION TROUBLESHOOTING

Indicators/Observations	Possible Cause	Check or Monitor	Possible Solutions
1. High solids in filtrate.	1a. Improper coagulant dosage.	1a. Coagulant dosage.	1a. Change coagulant dosage.
	1b. Filter media blinding due to polymer overfeed.	1b. Coagulant feeder calibration.	1b. Recalibrate coagulant feeder.
	1c. Filter media dirty.	1c. Visually inspect media.	1c. For synthetic cloth, use detergent and steam wash. For steel coils, acid clean clothwater wash or replace.
2. Thin cake with poor dewatering.	2a. Filter media blinding due to polymer overfeed.	2a. Inspect media and coagulant feed.	2a. Recalibrate coagulant feeder.
	2b. Improper chemical dosage.	2b. Coagulant dosage.	2b. Adjust coagulant dosage.
	2c. Inadequate vacuum.	2c. Amount of vacuum, leaks in vacuum system, leaks in seals.	2c. Repair vacuum system. See 3 below.
	2d. Drum speed too high.	2d. Drum speed.	2d. Increase drum speed.
	2e. Drum submergence too low.	2e. Drum submergence.	2e. Increase drum submergence.

VACUUM FILTRATION

Indicators/Observations	Possible Cause	Check or Monitor	Possible Solutions
3. Vacuum pump stops.	3a. Lack of power.	3a. Heater tripped.	3a. Reset pump switch.
	3b. Lack of seal water.	3b. Check seal water source.	3b. Reestablish seal water flow.
	3c. Broken V-belt.	3c. V-belt condition.	3c. Replace V-belt.
4. Drum stops rotating.	4. Lack of power.	4. Heater tripped.	4. Reset drum rotation switch.
5. Receiver is vibrating.	5a. Filtrate pump is clogged.	5a. Filtrate pump output.	5a. Turn pump off and clean.
	5b. Loose bolts and gasket around inspection plate.	5b. Inspection plate.	5b. Tighten bolts and align gasket.
	5c. Worn ball check in filtrate pump.	5c. Ball check.	5c. Replace ball check.
	5d. Air leaks in suction line.	5d. Suction line.	5d. Repair leaks.
	5e. Dirty drum face.	5e. Drum face.	5e. Clean face with pressure hose.
	5f. Seal strips missing.	5f. Drum.	5f. Replace seal strips.

Indicators/Observations	Possible Cause	Check or Monitor	Possible Solutions
6. Vat level too high.	6a. Improper chemical conditioning.	6a. Coagulant dosage.	6a. Adjust coagulant dosage.
	6b. Feed rate too high.	6b. Feed rate and solids yield.	6b. Reduce feed rate.
	6c. Drum speed too slow.	6c. Drum speed.	6c. Increase drum speed.
	6d. Filtrate pump off or clogged.	6d. Filtrate pump.	6d. Turn on or clean pump.
	6e. Drain line plugged.	6e. Drain line.	6e. Clean drain line.
	6f. Vacuum pump stopped.	6f. See 3 above.	6f. See 3 above.
	6g. Seal strips missing.	6g. Drum.	6g. Replace seal strips.
7. Vat level too low.	7a. Feed rate too low.	7a. Feed rate.	7a. Increase feed rate.
	7b. Vat drain valve open.	7b. Vat drain valve open.	7b. Close vat drain valve.

VACUUM FILTRATION

Indicators/Observations	Possible Cause	Check or Monitor	Possible Solutions
8. High amperage draw by vacuum pump.	8a. Filtrate pump clogged.	8a. Filtrate pump output.	8a. Turn pump off and clean.
	8b. Improper chemical conditioning.	8b. Coagulant dosage.	8b. Adjust coagulant dosage.
	8c. Vat level high.	8c. See 6 above.	8c. See 6 above.
	8d. Cooling water flow to vacuum pump too high.	8d. Cooling water flow.	8d. Decrease cooling water flow.
9. Scale buildup on vacuum pump seals.	9. Hard water.	9. Vacuum pump seals.	9. Add a scale inhibitor.

NOTES

23 PRESSURE FILTRATION

23.1. PROCESS OVERVIEW

- There are several types of filter presses available but the most common type consists of vertical plates that are held in a frame and are pressed together between a fixed and moving end.

- A cloth filter medium is mounted on the face of each individual plate.

- The press is closed and sludge is pumped into the press at pressures up to 225 psi and passes through feed holes in the trays along the length of the press.

- Filter presses usually require a precoat material such as incinerator ash or diatomaceous earth to aid in solids retention on the cloth and to allow easier release of the cake.

23.2. PRESSURE FILTRATION TROUBLESHOOTING

Indicators/Observations	Possible Cause	Check or Monitor	Possible Solutions
1. Plates fail to seal.	1a. Plates not aligned.	1a. Alignment.	1a. Realign plates.
	1b. Inadequate shimming. (Plates not level).	1b. Level.	1b. Adjust shimming to level unit.
2. Cake release is difficult.	2a. Inadequate precoat.	2a. Precoat feed.	2a. Increase precoat feed.
	2b. Improper conditioning.	2b. Conditioner type and dosage. Run filter leaf test.	2b. Change conditioner type or dosage based on filter leaf tests.
3. Excessive filter cycle times.	3a. Improper conditioning.	3a. Chemical dosage.	3a. Change chemical dosage.
	3b. Feed solids too low.	3b. Operation of thickening processes.	3b. Improve solids thickening to increase solids concentration in press feed.
4. Filter cake sticks to solids conveying equipment.	4. Change chemical conditioning by using more inorganic chemicals.	4. Conditioning dosage.	4. Decrease ash, increase inorganic conditioners.

PRESSURE FILTRATION

Indicators/Observations	Possible Cause	Check or Monitor	Possible Solutions
5. Precoat pressures gradually increase.	5a. Improper sludge conditioning.	5a. Conditioning dosages.	5a. Change chemical dosage.
	5b. Improper precoat feed.	5b. Precoat feed.	5b. Decrease precoat feed for a few cycles, then adjust to optimal amount.
	5c. Filter media plugged.	5c. Filter media.	5c. Wash filter media.
	5d. Calcium buildup in media.	5d. Filter media.	5d. Acid wash media (inhibited muriatic acid).
6. Frequent media blinding.	6a. Precoat inadequate.	6a. Precoat feed.	6a. Increase precoat.
	6b. Initial feed rates too high (where no precoat used).	6b. Feed rates.	6b. Develop initial cake slowly.
7. Excessive moisture in cake.	7a. Improper conditioning.	7a. Conditioning dosage.	7a. Change chemical dosage.
	7b. Filter cycle too short.	7b. Compare filtrate flow rate with cake moisture content.	7b. Lengthen filter cycle.

178 WASTEWATER TREATMENT TROUBLESHOOTING AND PROBLEM SOLVING

Indicators/Observations	Possible Cause	Check or Monitor	Possible Solutions
8. Sludge blowing out of press.	8. Obstruction, such as rags, in the press forcing sludge between plates.	8. Inspect press for presence of foreign material.	8. Shutdown feed pump, hit press closure drive, restart feed pump, clean feed eyes of plates at end of cycle.
9. Leaks around lower faces of plates.	9. Excessive wet cake soiling the media on lower faces.	9. Cake moisture content.	9. See 7 above.

NOTES

24 BELT FILTRATION

24.1. PROCESS OVERVIEW

- Belt filter presses avoid the sludge pick-up problem occasionally experienced with rotary vacuum filters.

- The sludge is gravity thickened, chemically conditioned with polymer and then pressure dewatered in the filter press to increase the solids content of either digested or undigested sludge.

- The influent mixture of sludge and polymer is fed to the moving porous belt on the filter press. Dewatering occurs as the sludge moves through a series of rollers which squeeze the sludge to the belt or squeeze the sludge between two belts.

- The cake is discharged from the belt by a scraper mechanism and the water is returned to the plant for treatment.

24.2. BELT FILTRATION TROUBLESHOOTING

Indicators/Observations	Possible Cause	Check or Monitor	Possible Solutions
1. Dewatered sludge not thick enough.	1a. Sludge application rate too high.	1a. Check sludge pumping rate.	1a. Adjust influent sludge pumping rate.
	1b. Belt speed too high.	1b. Check belt speed.	1b. Adjust belt speed.
	1c. Incorrect polymer dose.	1c. Check polymer mixing and dosage. Run a jar test.	1c. Use jar test procedure to determine optimum dosage.
2. Excessive belt wear.	2a. Improper alignment of rollers.	2a. Check tracking of belt to see if it creeps to one side.	2a. Adjust alignment of rollers.
	2b. Sludge buildup on bottom of belt on rollers causing improper alignment.	2b. Check operation of automatic belt adjustor. Check bottom of belt.	2b. Replace or repair faulty adjustor mechanism.
3. Solids in filtrate.	3a. Incorrect polymer dose.	3a. Check polymer mixing and dosage.	3a. Use jar test to determine optimum dosage.
	3b. Solids running off the edge of the filter belt.	3b. Check influent sludge pumping rate and belt rate of travel.	3b. Reduce sludge pumping rate accordingly. Adjust belt rate of travel as required.

BELT FILTRATION

Indicators/Observations	Possible Cause	Check or Monitor	Possible Solutions
4. Oil leak.	4. Oil seal failure.	4. Check oil seal.	4. Replace seal.
5. Noisy or hot bearings or universal joint.	5a. Excessive wear due to improper alignment.	5a. Alignment.	5a. Replace, lubricate, or align joint or bearing as required.
	5b. Lack of lubrication.	5b. Lubrication.	5b. Remove excess lubrication and lubricate according to manufacturer's recommendations.

NOTES

25 BASIC MECHANICAL PROBLEMS

- The following troubleshooting table describes problems and possible corrective measures for mechanical failures that could be encountered in any of the previous process sections.

25.1. BASIC MECHANICAL TROUBLESHOOTING

Indicators/Observations	Possible Cause		Check or Monitor		Possible Solutions	
1. Frequent chain or shear pin failure.	1a.	Wrong size shear pin.	1a.	Shear pin size.	1a.	Change shear pin size.
	1b.	Flights not aligned.	1b.	Alignment.	1b.	Realign if necessary.
	1c.	Ice.	1c.	Walls and surfaces for ice buildup.	1c.	Remove or break up ice.
	1d.	Excessive sludge.	1d.	Sludge levels.	1d1.	Operate collectors more often and for longer periods of time.
					1d2.	Remove sludge more often.
2. Broken chain or sprockets.	2a.	Wrong size chain or sprocket.	2a.	Chain and sprocket size.	2a.	Replace with correct size.
	2b.	Excessive hydraulic shock load.	2b.	Check for hydraulic surges.	2b.	Install inlet baffle to deflect force.
	2c.	Misalignment.	2c.	Alignment.	2c.	Correct alignment.

BASIC MECHANICAL PROBLEMS 187

Indicators/Observations	Possible Cause	Check or Monitor	Possible Solutions
3. Noisy drive.	3a. Loose chain.	3a. Chain tension.	3a. Take up slack (remove one or more chain links).
	3b. Chain and/or sprockets are worn.	3b. Equipment wear.	3b. Replace or remove equipment.
	3c. Chain does not fit sprockets.	3c. Chain and sprocket size.	3c. Replace with correct size equipment.
	3d. Worn bearings.	3d. Bearings.	3d. Replace if necessary.
	3e. Improper lubrication.	3e. Lubrication requirements.	3e. Check manufacturer's literature and re-lubricate if necessary.
	3f. Circular tank central drive mechanism is worn.	3f. Alignment.	3f. Realign.
	3g. Circular tank central drive mechanism is worn.	3g. Drive wear.	3g. Check bearings and lubrication. Repair, replace, or service as necessary.

Indicators/Observations	Possible Cause	Check or Monitor	Possible Solutions
4. Stiff chain.	4a. Chain is worn.	4a. Chain wear.	4a. Replace chain.
	4b. Improper lubrication.	4b. Lubrication requirement.	4b. Lubricate properly.
	4c. Rust, dirt, or corrosion.	4c. Equipment appearance.	4c. Clean and lubricate.
	4d. Chain misalignment.	4d. Alignment.	4d. Realign chain and sprocket.
5. Rapid chain wear.	5a. Loose or misaligned chain.	5a. Alignment.	5a. Tighten and realign as necessary.
	5b. Incorrect lubrication.	5b. Lubrication requirement.	5b. Lubricate properly.
6. Chain "climbs" sprocket.	6a. Worn chain or sprockets.	6a. Chain and sprocket wear.	6a. Replace worn equipment.
	6b. Chain does not fit sprockets.	6b. Chain and sprocket size.	6b. Replace with correct size equipment.
	6c. Loose chain.	6c. Chain tension.	6c. Tighten chain (remove one or more links).
	6d. Misalignment.	6d. Alignment.	6d. Align chain and sprocket.

BASIC MECHANICAL PROBLEMS 189

Indicators/Observations	Possible Cause	Check or Monitor	Possible Solutions
7. Corrosion of equipment.	7a. Septic wastewater.	7a. Color and odor of wastewater.	7a1. Repaint surfaces with corrosion resistant coatings. 7a2. Investigate and correct cause or source of septic wastewater.
8. Leaking oil seals.	8. Oil seal failure.	8. Oil seal.	8. Replace seal.
9. Pump malfunction.	9a. Pump blockage.	9a. Check for trash in pump.	9a. Remove blockage from pump.
	9b. High discharge pressure.	9b. Check for closed valve or blockage on discharge side of pump.	9b. Remove blockage.
	9c. Pump overload.	9c. Load limits.	9c. Remove pump from service and start another pump, or try extra pumps.

Indicators/Observations	Possible Cause	Check or Monitor	Possible Solutions
10. Pump not delivering	10a. Pump not primed.	10a. Prime.	10a. Allow sludge to fill pump to ensure prime.
	10b. Air in casing.	10b. Casing.	10b. Bleed air from casing and restart the pump.
	10c. Closed suction valve.	10c. Valve.	10c. Open valve.
	10d. Closed discharge valve.	10d. Valve.	10d. Open valve.
	10e. Impeller clogged.	10e. Impeller.	10e. Clean impeller.
	10f. Wrong direction of rotation.	10f. Shaft rotation.	10f. Check rotation and correct if necessary.
	10g. Sludge lines plugged.	10g. Sludge lines.	10g. Clean lines.

BASIC MECHANICAL PROBLEMS   191

Indicators/Observations	Possible Cause	Check or Monitor	Possible Solutions
11. Vibration and/or noise.	11a. Impeller clogged.	11a. Impeller.	11a. Clean impeller.
	11b. Damaged impeller.	11b. Impeller.	11b. Replace impeller.
	11c. Improper bearing lubrication.	11c. Check lubrication requirements.	11c. Use proper lubricant and drain any excess.
	11d. Pump out of alignment.	11d. Alignment.	11d. Realign if necessary.
	11e. Worn bearings.	11e. Bearings.	11e. Replace if necessary.
	11f. Bent drive shaft	11f. Shaft.	11f. Replace if necessary.
	11g. Cause not apparent.	11g. Scratch your head.	11g. Remove pump from service until cause is determined and corrected.
12. Noisy or hot bearings.	12a. Excessive wear.	12a. Bearings.	12a. Replace bearings.
	12b. Improper lubrication.	12b. Check lubricant and frequency requirements.	12b. Lubricate or remove excess.

Indicators/Observations	Possible Cause	Check or Monitor	Possible Solutions
13. Motor runs hot.	13a. Discharge valve closed.	13a. Check valve position.	13a. Open valve.
	13b. Speed too high.	13b. Check speed.	13b. Reduce motor speed.
	13c. Impeller clogged.	13c. Impeller.	13c. Clean impeller.
	13d. Defective motor.	13d. Motor input and output.	13d. Replace or repair motor.
	13e. Incorrect voltage.	13e. Motor voltage.	13e. Adjust voltage or replace motor.
	13f. Incorrect lubrication.	13f. Check lubricant and frequency requirements.	13f. Lubricate or remove excess.
	13g. Pumped liquid too thick.	13g. Check viscosity of liquid.	13g1. Dilute with flush water.
			13g2. Use larger pump.
			13g3. Pump sludge more frequently.
	13h. Impeller closed.	13h. Check for debris around impeller.	13h. Clean impeller or flush with clean water.

BASIC MECHANICAL PROBLEMS 193

Indicators/Observations	Possible Cause	Check or Monitor	Possible Solutions
14. Excessive pump packing leakage.	14. Packing requires adjustment or replacement.	14. Packing leakage.	14a. Adjust packing by tightening packing gland. 14b. Remove pump from service and replace packing.
15. Seal water pressure low.	15a. Seal water line has broken.	15a. Seal water line.	15a. Repair line.
	15b. Pump packing is too loose and seal water is flowing from pump.	15b. Packing leakage.	15b. Tighten or replace pump packing.
	15c. Strainer is clogged.	15c. Strainer.	15c. Clean strainer.
	15d. Solenoid valve did not open.	15d. Solenoid valve. Open bypass to check pressure at pump.	15d. Replace solenoid if necessary.

WASTEWATER TREATMENT TROUBLESHOOTING AND PROBLEM SOLVING

Indicators/Observations	Possible Cause	Check or Monitor	Possible Solutions
16. Pump draws too much power.	16a. Pumping too much liquid.	16a. Check head vs. pump curve.	16a. Machine impeller OD to size advised by manufacturer.
	16b. Cavitation.	16b. Check lift vs. pump curve.	16b. Reduce suction lift.
	16c. Shaft bent or misaligned.	16c. Check deflection (0.002").	16c. Replace shaft or re-align.
	16d. Speed too high.	16d. Speed control.	16d. Decrease speed.
	16e. Wrong rotation.	16e. Check rotation with arrow on pump.	16e. Swap any two pump leads.
17. Seal leaking.	17a. Faces stuck open.	17a. Solids embedded in face or elastomer sticks to shaft.	17a. Replace face. Replace O-ring.
	17b. Excessive heat.	17b. Elastomer shape. Corrosion.	17b. Replace O-rings. Replace parts.
	17c. Incorrect installation.	17c. Loose gland. Misalignment.	17c. Tighten evenly.

BASIC MECHANICAL PROBLEMS 195

Indicators/Observations	Possible Cause	Check or Monitor	Possible Solutions
18. Pump operates for short time, then stops delivering.	18a. Incomplete priming.	18a. Air in pump or pipes.	18a. Bleed air from pump, piping, and valves.
	18b. Air leak in suction.	18b. Leaks in suction pipe.	18b. Vent volute casing of air and fill with liquid.
	18c. Air in liquid.	18c. Look for aeration piping near suction piping.	18c. Decrease air to wet well. Move aeration piping away from suction piping.
19. Blower problems. No air flow.	19a. Speed too low.	19a. Check motor and speed.	19a. Have electrician check voltage.
	19b. Wrong rotation.	19b. Check direction of rotation.	19b. Correct rotation direction.
	19c. Obstruction in piping.	19c. Check valve positions and for debris in valves, piping, silencers, or screens.	19c. Remove debris and/or open closed valves.

Indicators/Observations	Possible Cause	Check or Monitor	Possible Solutions
20. Blower problems. Low air flow.	20a. Speed too low.	20a. Check motor and speed.	20a. Have electrician check voltage.
	20b. Excessive pressure.	20b. Check inlet vacuum and discharge pressure.	20b. Reset relief valve.
	20c. Excessive slippage.	20c. Check inside casing for eroded surfaces.	20c. Restore to proper clearances.
	20d. Obstruction in piping or closed valve.	20d. Check valve positions and for debris in valves, piping, silencers, or screens.	20d. Remove debris and/or open closed valves.
21. Blower vibration.	21a. Improper coupling alignment.	21a. Check alignment.	21a. Realign.
	21b. Impeller rubbing.	21b. Check impeller clearances.	21b. Adjust impeller to stop rubbing.
	21c. Worn bearings or gears.	21c. Check bearings and gears.	21c. Replace worn components.
	21d. Unbalanced or rubbing impellers.	21d. Check for scale buildup on casing or impellers.	21d. Remove scale to restore clearances.
	21e. Drive or blower loose.	21e. Check mounting bolts.	21e. Tighten bolts.

BASIC MECHANICAL PROBLEMS 197

Indicators/Observations	Possible Cause	Check or Monitor	Possible Solutions
22. Worn side patterns on V-belt.	22a. Constant slip.	22a. Shriek at start-up.	22a. Retension drive until belt stops slipping.
	22b. Misalignment.	22b. Check alignment.	22b. Realign sheaves.
	22c. Worn sheaves.	22c. Check sheave wear.	22c. Replace with new sheaves.
	22d. Incorrect belt.	22d. Check belt size.	22d. Replace with correct belt.
23. Bottom of belt cracked.	23a. Belt slippage causing heat buildup and gradual hardening of undercord.	23a. Check for slippage.	23a. Install new belt and tension correctly to prevent slippage.
	23b. Idler installed on wrong side of belt.	23b. Check idler position.	23b. Refer to your V-belt installation manual.
	23c. Improper storage (dry rot).	23c. Check storage practices.	23c. Refer to manufacturer's belt storage guidelines.
24. V-belt bottom and sides burned.	24a. Belt slipping under starting load.	24a. Listen for belt shriek at start-up.	24a. Replace belt and tighten drive until slipping stops.
	24b. Worn sheaves.	24b. Check sheave wear.	24b. Replace sheaves.

Indicators/Observations	Possible Cause	Check or Monitor	Possible Solutions
25. V-belt turns over.	25a. Foreign material in grooves.	25a. Check grooves.	25a. Remove material and shield drive.
	25b. Misaligned sheaves.	25b. Check sheave alignment.	25b. Realign the drive.
	25c. Worn sheave grooves.	25c. Check sheave grooves.	25c. Replace sheave.
	25d. Tensile member broken by improper installation.	25d. Check tensile member.	25d. Replace with new belt(s).
	25e. Incorrectly aligned idler pulley.	25e. Check idler pulley alignment.	25e. Carefully align idler, checking alignment with drive loaded and unloaded.
26. V-belt pulled apart.	26a. Extreme shock load.	26a. Check start-up load.	26a. Remove cause of shock load.
	26b. Belt came off drive.	26b. Observe belt position.	26b. Check drive alignment, foreign material in drive, and ensure proper tension and drive alignment.

NOTES

BIBLIOGRAPHY

Adams, V.D., 1990. *Water and Wastewater Examination Manual* (Chelsea, MI: Lewis Publishers).

Bloch, H.P. and F.K. Geitner, 1983. *Machinery Failure Analysis and Troubleshooting* (Houston: Gulf Publishing Company).

Bloch, H.B., 1989. *Process Plant Machinery* (Stoneham, MA: Butterworth Publishers).

Cameron, W. and F.L. Cross, Jr., 1976. *Operation and Maintenance of Sewage Treatment Plants* (Westport, CT: Technomic Publishing Company).

Degremont [Company], 1979. *Water Treatment Handbook*, Fifth Edition (New York: Halsted Press).

Elonka, S.M., 1980. *Standard Plant Operator's Manual*, Third Edition (New York: McGraw-Hill Book Company).

Green, J.H. and A. Kramer, 1979. *Food Processing Waste Management* (Westport, CT: AVI Publishing Company).

Hammer, H., 1988. *Maintenance Mechanic* (New York: Prentice Hall Press).

Hicks, Tyler G., 1958. *Pump Operation and Maintenance* (New York: McGraw-Hill Book Company).

Kerri, K.D., editor, 1988. *Operation of Wastewater Treatment Plants*, Volumes 1 and 2 (Sacramento, CA: Hornet Foundation).

Kerri, K.D., editor, 1988. *Advanced Waste Treatment* (Sacramento, CA: Hornet Foundation).

Lansdown, A.R., 1982. *Lubrication, A Practical Guide to Lubricant Selection* (Elmsford, NY: Pergamon Press).

Martin, E.J. and E.T. Martin, 1991. *Technologies for Small Water and Wastewater Systems* (New York: Van Nostrand Reinhold).

Ministry of the Environment, 1984. *Preventive Maintenance Workshop*, Toronto, Ontario.

Nelson, C.A., 1981. *Mechanical Trades Pocket Manual* (Toronto: Forum House Publishing Company).

Rosaler, R.C., J.O. Rice, and T.G. Hicks, Editors, 1983. *Plant Equipment Reference Guide*, (New York: McGraw-Hill Book Company).

Tillman, G.M., 1992. *Primary Treatment at Wastewater Treatment Plants* (Chelsea, MI: Lewis Publishers).

Tillman, G.M., 1991. *Basic Mechanical Maintenance Procedures at Water and Wastewater Plants* (Chelsea, MI: Lewis Publishers).

U.S. Environmental Protection Agency, 1978. *Field Manual for Performance Evaluations and Troubleshooting at Municipal Wastewater Treatment Facilities.*

U.S. Environmental Protection Agency, 1978. *Operations Manual Sludge Handling and Conditioning.*

Water Pollution Control Federation, 1990. *Operation of Municipal Wastewater Treatment Plants, Manual of Practice No. II, Volumes II and III*, Second Edition, Alexandria, Virginia.